T0310201

Big Data and Differential Privacy

Wiley Series in Operations Research and Management Science

Wiley Series in Operations Research and Management Science

A complete list of the titles in this series appears at the end of this volume.

Big Data and Differential Privacy

Big Data and Differential Privacy

Analysis Strategies for Railway Track Engineering

Nii O. Attoh-Okine

This edition first published 2017
©2017 John Wiley & Sons, Inc.

The right of Nii O. Attoh-Okine to be identified as the author of this work has been asserted in accordance with law.

Registered Offices
John Wiley & Sons, Inc., 111 River Street, Hoboken, NJ 07030, USA

Editorial Office
111 River Street, Hoboken, NJ 07030, USA

For details of our global editorial offices, customer services, and more information about Wiley products visit us at www.wiley.com.

Wiley also publishes its books in a variety of electronic formats and by print-on-demand. Some content that appears in standard print versions of this book may not be available in other formats.

Library of Congress Cataloguing-in-Publication Data

Names: Attoh-Okine, Nii O., author.
Title: Big data and differential privacy : analysis strategies for railway
    track engineering / Nii O. Attoh-Okine.
Other titles: Wiley series in operations research and management science.
Description: Hoboken, NJ : John Wiley & Sons, 2017. | Series: Wiley series in
    operations research and management science | Includes bibliographical
    references and index.
Identifiers: LCCN 2017005398 (print) | LCCN 2017010092 (ebook) | ISBN
    9781119229049 (cloth) | ISBN 9781119229056 (pdf) | ISBN 9781119229063
    (epub)
Subjects: LCSH: Railroad tracks–Mathematical models. | Data
    protection–Mathematics. | Big data. | Differential equations.
Classification: LCC TF241 .A88 2017 (print) | LCC TF241 (ebook) | DDC
    625.1/4028557–dc23
LC record available at https://lccn.loc.gov/2017005398

Cover design: Wiley
Cover image: (Top Image) © Jaap Hart/iStockphoto; (Bottom Image) © mbbirdy/Gettyimages

Set in 10/12pt WarnockPro by SPi Global, Chennai, India

Printed in the United States of America

10 9 8 7 6 5 4 3 2 1

Contents

Preface

The ability of railway track engineers to handle and process large and continuous streams of data will provide a considerable opportunity for railway agencies. This will help decision makers to make informed decisions about the maintenance, reliability, and safety of the railway tracks. Now a period is beginning in which the problem is collecting the railway track data and analyzing it in a defined period of time. Therefore, the tools and methods needed to achieve this analysis need to be addressed. Knowledge derived from big data analytics in railway track engineering will become one of the foundational elements of any railway organization and agency. Also, another key issue has been the protection of data by different railway organizations. Therefore, although the data are available, they are really shared among different agencies. This makes the issue of differential privacy of utmost importance in the railway industry. Also, it is not clear if the industry has developed a clear way of both protecting and accessing the data from third parties.

Data science is an emerging field that has all the characteristics needed by railway track engineers to address and handle the enormous amounts of data generated by various technology platforms currently in place. The major objective is for railway track engineers to have an understanding of big data. Using the right tools and methodologies, railway track big data will also uncover new directions for monitoring and collecting railway track data; this apart from the engineering side will also have a major business impact on railway agencies.

This book provides the fundamental concepts needed to work with big data applications for railway engineers. The concepts serve as a foundation, and it is assumed that the reader has some understanding of railway engineering. The book does not attempt to address railway track engineering as a subject, but it does address the use of data science and the big data paradigm in railway track applications. Colleagues in industry will find the book very handy, but it will also serve as a new direction for graduate students interested in data science and the big data paradigm in infrastructure systems. The work in this book is intended to be accessible to an audience broader than those in railway track engineering.

Furthermore, I hope to shed a bright light on the enormous potential and future development that the big data paradigm will bring to railway track engineering. The amount of data railway agencies already have and the amount they are planning to collect in the future make this book an important milestone. This book attempts to bring together new emerging topics in a coherent way that can address different methodologies that can be used in solving a variety of railway track problems in the analysis of large data from various inspection technologies. In preparing the book, I tried to achieve the following objectives: (a) to develop some data science ontologies, (b) to provide the formulation of large railway track data using big data analytics, (c) to provide direction on how to present the data (visualization of the results), (d) to provide practical applications for the railway and infrastructure industry, and (e) to provide a new direction in railway track data analysis.

Finally, I assume full responsibility for any errors in the book. The opinions presented in the book represent my experiences in civil infrastructure systems, machine learning, signal analysis, and probability analysis.

January, 2016

Nii O. Attoh-Okine
Newark, Delaware, USA

Acknowledgments

I would like to thank the staff of John Wiley & Sons, Inc., especially Susanne Steitz-Filler, for their time. I would also like to thank Dr. Allan Zarembski and Joe Palese and Hugh Thompson of FRA for their support and encouragement. Thanks also to my current and former graduate students Dr. Yaw Adu-Gyamfi, Dr. Offei Adarkwa, and Emmanuel Martey for offering constructive criticisms. Special thanks to Silvia Galvan-Nunez who additionally provided me support with the complex LaTex issues. I would also like to thank Erin Huston for editing the first draft of the book. Finally, as always, I would like to thank my family: my two children, Nii Attoh and Naa Djama; my wife, Rebecca, for providing the peace and excellent working environment; and my brother, Ashalley Attoh-Okine, an excellent actuary and energy expert, who introduced me to so many data analysis techniques, which have been part of my research over the years. I dedicate the book to the memory of my parents, Madam Charkor Quaynor and Richard Ayi Attoh-Okine, and my maternal grandparents, Madam Botor Clottey and Robert Quaynor.

1

Introduction

1.1 General

Currently, railroads collect enormous quantities of data through vehicle-based inspection cars, trackside (or wayside) monitoring systems, hand-held gauges, and visual inspections. In addition, these data are located geographically using the global positioning system (GPS). The data from these inspection systems are collected electronically by hand or using various sensors, video inspections, machine visions, and many other sources. Furthermore, the data are growing both in quantity and quality and are more precise and diverse. Data of extremely large sizes are difficult to analyze using traditional approaches since they may exceed the limits of a typical spreadsheet. The railway track data are present in diverse forms, including categorical, numerical, or continuous values. The general characteristics of the data dictate which type of method is appropriate for analysis. For example, categorical and nominal values are unsorted, while numerical and continuous values are assumed to be sorted or to represent ordinal data (Ramírez-Gallego et al., 2016).

The development of advanced sensors and information technology in railway infrastructure monitoring and control has provided a platform for the expansive growth of data. This has created a new paradigm in the processing, storing, streaming, and visualization of data and information. Furthermore, changes in technology include the possibility of installing sensors and smart chips in critical infrastructure to measure system performance, current condition, and other indicators of imminent failures. Many of the railway infrastructure components have communication capabilities that allow data to be uploaded on demand.

Big data is about extremely large volumes of data originating from various sources: databases, audio and video, millions of sensors, and other systems. The sources of data in some cases provide structured outputs, but most are unstructured, semi-structured, or poly-structured. These data are streaming in some cases with high velocity, and the data exposes at a higher speed or some speed as it is generated.

Big Data and Differential Privacy: Analysis Strategies for Railway Track Engineering, First Edition. Nii O. Attoh-Okine.
© 2017 John Wiley & Sons, Inc. Published 2017 by John Wiley & Sons, Inc.

This chapter presents a general overview, basic description, and properties of deterministic and random data that are encountered in railway track engineering data and relies heavily on the data output based on the advances in sensors, information technology, high information technology, and development that has led to extremely massive data sets. These large data sets have made the traditional analytical techniques used for railway track maintenance and safety issues somewhat obsolete.

The data obtained in railway track monitoring are collected by different sensors, at different times and environmental conditions, at different frequencies, and at different resolutions. The outputs of these data have different characteristics: discrete or continuous, spatial or temporal, signal and images, and categorical and objective, among others. All these characteristics, properties, and the extreme volume of data collected have made traditional analytical techniques very inefficient; issues like visualization and data streaming, which are very critical in railway track maintenance and safety, are not adequately addressed. The traditional statistical techniques fail to scale up to the extremely large volumes of data collected by railway inspection vehicles and trackside monitoring devices. Therefore, the growing amount of data generated by railway track inspection activities is outpacing the current capacity to explore and interpret these data and hence appropriately addresses maintenance and safety issues.

1.2   Track Components

The term "tracks" includes superstructure, substructure, and special structures (Figure 1.1). The superstructure is made of rails, ties, fasteners, turnouts, and crossings, while the substructure consists of ballast, subballast, the subgrade, and other drainage facilities. The superstructure and substructure are separated by the tie–ballast interface.

The main purpose of the railway track structure is to provide a safe and economical train transportation system through guiding the vehicle and transmitting loads through the track components to the subgrade. The carrying

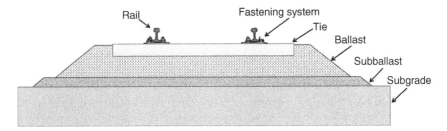

Figure 1.1 Track structure components

capacity and long-term durability of the track structure highly depend on how the superstructure and substructure respond to and interact with each other when subjected to moving trains and environmental factors (Selig and Waters, 1994; Kerr, 2003).

The function of different rail components has been presented by various authors, such as Hay (1982), Selig and Waters (1994), Esveld (2001), Kerr (2003), Sadeghi (2010), and Tzanakakis (2013). The aim of this section is to summarize this function. The rails are the longitudinal steel members that are placed on spaced ties to guide the train wheels evenly and continuously. Their strength and stiffness must be sufficient to maintain a steady shape and smooth track configuration and to resist various forces (vertical, lateral, and longitudinal) by vehicles. The rails also in some cases serve as electrical conductors for the signal circuit and also as a groundline for the electric locomotive power circuit. The profile of the rail surface (transverse and longitudinal) and wheel surface has a major influence on the operation of the vehicles on the track, and track defects may in some instances create and cause large dynamic loads that lead to derailment and safety issues, as well as accelerated degradation.

Most steel rail sections are connected either by bolted joints or by welding. The bolted joints create several problems, including rough riding track, undesirable vibration, and additional impact loads, among others; hence, the use of continuous welded rail (CWR) has been the better solution. CWR attempts to address some of the disadvantages of the bolted joints, which have its own set of maintenance requirements.

The rail fastener systems, or fastenings, include all the components that connect the rail to the tie, with the tie plate, spike, and anchor for wood ties and clip, insulator, and elastic fasteners for concrete ties. The function of the fastenings is to retain the rail against the ties and resist vertical, lateral, longitudinal, and overturning movements of the rail. They also serve as wheel load impact attenuation, increasing track elasticity, as well as electrical isolation between rails.

For concrete tie tracks, rail pads are installed on rail supporting points to reduce and transfer the stress and dynamic forces from the rail to the ties, and they reduce the interaction force between the rail and the ties (Choi, 2011). The pads also provide adequate resistance to longitudinal and rotational movement of the rail and provide a conforming layer between the rail and tie to avoid contact areas of high pressure. From a dynamic point of view, the rail pads tend to influence overall track stiffness.

Ties are transverse beams resting on ballast and support. They span below and tie together two rails. The main functions of ties are as follows:

- Uniformly transfer and distribute loads from the rail to the ballast
- Hold the fastening system to maintain proper track gage
- Restrain the lateral, longitudinal, and vertical rail movement by anchorage of the superstructure to the ballast

- Provide a cant to the rails to help develop proper wheel–rail contact by matching the inclination of the conical wheel shape
- Provide an insulation layer
- Allow fast drainage of fluid
- Allow for proper ballast maintenance

Ballast is the layer of crushed stone placed at the top layer of the substructure in which the tie is embedded. It is an elastic support and transfers forces from the rail and tie to the subballast. As some of its functions, it

- Distributes load from ties uniformly over the subgrade
- Anchors the track in place against lateral, vertical, and longitudinal movements
- Absorbs shock from the dynamic load
- Allows suitable global and local track settlement
- Avoids freezing and melting (thawing) problems by frost action
- Allows for proper drainage
- Allows for maintenance of the track geometry

The subballast is the layer between the ballast and the subgrade. As some of its functions, it

- Reduces the stress at the bottom of the ballast layers to a reasonable level to protect the subgrade
- Migrates fines from the subgrade to the upper layer of the ballast
- Protects the subgrade from the ballast
- Permits drainage of water that might otherwise flow upward from the subgrade

The subgrade is the last support of the track systems and, in some cases, is the existing soil at the location, unless the existing formation is very weak. In the case of a weak existing formation, techniques like stabilization and modification of the existing elevation use more appropriate soil. The addition of geosynthetic material has been used to improve the subgrade performance and bearing capacity. Its main functions are the following:

- Provide support to the track structure
- Bear and distribute the resultant load from the train vehicle through the track structure
- Provide sufficient drainage

1.3 Characteristics of Railway Track Data

Railway track data are similar to data from other infrastructures. Its characteristics include the following:

- *Massive Data Sets.* Railway track data collection and monitoring has resulted in extremely large data sets for infrastructure monitoring. In some cases, the actual data are processed and only the reduced version is stored, while in most cases smaller amounts of data are stored for further analysis.
- *Unstructured Data, Heterogeneous Databases.* Some of the railway track data are stored in databases. In most cases, different agencies and countries have different data formats, different database management systems, and different data manipulation algorithms. Most of these databases are evolving, which in some cases makes analysis and data mining across them challenging. Some of the databases include unstructured images, plots, and tables, as well as links to other transportation and infrastructure documents of the agency. This can be challenging in terms of both analysis and reporting.
- *Information in the Form of Images.* The analysis of railway track, in terms of both rail and geometry defects, by its very nature deals with issues associated with the extraction of meaningful information from massive amounts of railway track images, thus opening a new direction in railway track analysis.
- *Poor Quality of Data.* Railway track data analysis, especially the image data, in most cases is of poor quality due to the railway track environment and sensor noise. In some cases, data are missing or input incorrectly. Furthermore, the data from different sources can vary in terms of quality. Also, the railway inspectors may in some cases have incomplete knowledge about the mechanism and initiation of different defects. This may lead to inconclusive reporting and analysis.
- *Multiresolution and Multisensor Data.* Several different sensors are used to collect different information and data. This may create a situation where several images may have different resolutions over time. Therefore, care must be taken so that the change in resolution can be included.
- *Noisy Data.* Noisy data cannot be avoided in railway track data collections. Methods of reducing the noise in data need to be implemented during the preprocessing of the data for further analysis. For example, shadows and orientations of the vehicle collecting the data can have an impact on the images. Therefore, poor illumination can have a major impact on the obtained image.
- *Missing Data.* The risk of missing data is always present in railway track data collection; this is mostly due to sensor malfunction. Filling the gaps can be a daunting task. Again care must be taken with how missing data is included.
- *Streaming Data.* Some of the data sets collected during railway monitoring can be streaming in nature; that is, a constant stream of data is being collected and received. This requires a specialized set of analyses different from the chunk data methods used in traditional analysis.

More broadly, the data can either be random or deterministic. The random data is shown in Figure 1.2, and the deterministic data is shown in Figure 1.3, as presented by Bendat (1998).

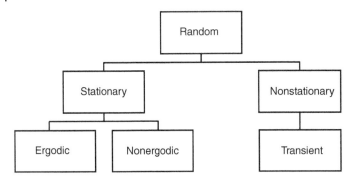

Figure 1.2 Classification of random data (Bendat (1998). Reproduced with their permission of John Wiley & Sons)

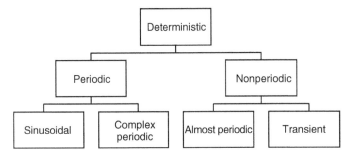

Figure 1.3 Classification of deterministic data (Bendat (1998). Reproduced with their permission of John Wiley & Sons)

Table 1.1 shows the general taxonomy of big data methods in railway engineering.

1.4 Railway Track Engineering Problems

Generally, railway track engineering problems can be classified into two groups according to Santamarina and Fratta (1998): (a) forward problems and (b) inverse problems. Table 1.2 shows the group of problems that fall under the two categories. For forward problems, the major objective is to design systems to satisfy predefined performance criteria. Also, convolution forms part of the forward problems. In convolution, the input is known, the type of system is known, and the only unknown is the output.

Inverse problems can either be (a) system identification where the input and output are known but the system is unknown or (b) deconvolution where the input is unknown, while the system and output are known. Figure 1.4 shows a generic representation of general civil engineering problems, including railway

Table 1.1 Taxonomy of big data in railway engineering.

Analysis domain	Sources	Characteristics	Approaches	Comments
Structured data	Field data collection, sensors, data from scientific experiments	Structured records, real time	Data mining, statistical analysis	All infrastructure systems need field data
Unstructured data	Extreme events, sensors	Unstructured records, mixture of variables	Anomaly detection	Infrastructure inspection reports, specification updates
Text analytics	Logs, email, corporate documents, government rules and regulations, text content of web pages, citizen feedback and comments	Unstructured, rich textual, context, semantic, language dependent	Document presentation, NLP, information extraction, topic model, summarization, categorization, clustering, question answering, option mining	Early detection
Multimedia analytics	Corporation-produced multimedia, user-generated multimedia, surveillance	Image, audio, video, massive, redundancy, semantic gap	Summarization, annotation, indexing and retrieval, recommendation, event detection	Early detection
Mobile analytics	Mobile apps, RFID sensors		Monitoring, location-based mining	

Table 1.2 Engineering problems.

Forward problems		Inverse problems	
System design	Convolution	System identification	Deconvolution
The system is designed to satisfy performance criteria (controlled output for estimated input)	Input: known System: known Output: unknown	Input: known System: unknown Output: known	Input: unknown System: known Output: known

(Santamarina and Fratta., 1998) Reproduced with the permission of ASCE.

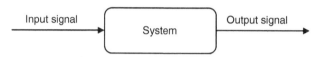

Figure 1.4 Engineering signals

track problems. But there can be different situations, including (a) single-input–output relationships as shown on the generic representation and (b) multiple-input–output relationships. Therefore, depending on the structural and objective analyses, there are different assumptions and analyses. Systems can be divided into two broad groups, (a) linear and (b) nonlinear. The linear systems can be further divided into constant parameter and time-varying systems (Bendat, 1998).

Major parts of railway track data are in the form of signals and images; therefore, a deeper understanding of analytical issues for signals and images is needed to analyze and interpret railway track data. A major issue related to track images is the presence of noise, which tends to affect the overall images if it is not properly reduced or accounted for. Therefore, efficient algorithms are needed to reduce noise in railway track images before further analysis can be done.

Table 1.3 shows examples of different track inspection technologies and their level of maturity. Railway track conditions are, in most cases, evaluated using the characteristics of track geometry wave form and vehicle dynamic response to the track. Also, in some cases, images from high definition cameras are also collected. It is apparent that to obtain the true picture of the railway track condition, there should be methods that can go beyond traditional statistical analysis. An efficient method is one that can perform the mining of the data, reduce noise from the wave forms, and combine data and information from different sources to provide a clear understanding of what maintenance activities to perform and how to satisfy all safety requirements.

Table 1.3 Track inspection technologies.

Component	Inspection	Base technology	Maturity	Measurement	Maintenance activity	Level of automation
Rail	Profile	Laser/camera	Mature	Wear, cant, lip, GF angle, profile/contact	Wear, grinding, cant, lip	Very high
	Corrugation	Inertial or chord	Moderate	Wavelength and amplitude	Grinding	High
	Internal fatigue	Ultrasonic	Mature	Location and size of flaw	Replacement	High
	Surface fatigue	High definition camera	New	Location and size of flaw	Grinding	Moderate
	Surface cracks	Eddy current	New	Location, length, and density of cracks	Grinding	Low
Joint bars	Cracks, bolts	High definition camera	New	Location of failure	Replacement	Moderate
Ties	Cracks, plate cut	High definition camera	New	Length, width, and density of cracks	Replacement	Low
	Rot	Back scatter X-ray	Very new	Internal density	Replacement	Low
	GRMS	Force/displacement	Mature	Tie/fastener interface strength/stiffness	Replacement	High
Fasteners	Condition, existence	High definition camera	New	Missing or damaged	Replacement	Low
	GRMS	Force/displacement	Mature	Tie/fastener interface strength/stiffness	Replacement	High
Turnouts/ diamonds	Rail geometry	Laser/camera	New	Relative heights and dimensions, points, frog	Replacement, grinding, welding	Moderate
	Track geometry	Inertial	Mature	Relative rail-to-rail relationship	Tamping	High
	Component condition	High definition camera	New	Missing or damaged	Replacement, tighten	Low

(Continued)

Table 1.3 (Continued)

Component	Inspection	Base technology	Maturity	Measurement	Maintenance activity	Level of automation
Geometry	Geometry car	Inertial or contact	Mature	Relative rail-to-rail relationship	Tamping	High
	Absolute geometry	Inertial	New	Absolute rail-to-rail relationship	Tamping	Low
	Vertical track interaction	Inertial	New	Vertical vehicle response to track	Tamping	High
Ballast/ subballast/ subgrade	Ground-penetrating radar	Radar	Moderate	Ballast depth and condition	Tamping, undercutting	Low
	Cone penetrometer	Force/displacement	Moderate	Ballast/subballast depth, condition, and strength	Tamping, undercutting	Low
	Vertical track deflection	Force/displacement	New	Track stiffness	Tamping, undercutting	Moderate
	Profile	Lidar	New	Ballast profile	Add ballast and tamp	Moderate
Clearance	Envelope	Lidar/laser	Moderate	Surrounding clearance and obstructions		Moderate

1.5   Wheel–Rail Interface Data

The wheel–rail contacts at the interface between the wheel and rail determine in part the reliability of railway systems. Tzanakakis (2013) presented in Figure 1.5 the different outcomes and effects of wheel–rail contact. The rail vehicles are supported, accelerated, and decelerated by contact forces acting on extremely small wheel–rail contact areas (around 1 cm^2).

Meymand et al. (2016) presented a comprehensive survey on the topic. The paper discusses well-known theories for modeling normal contacts based on Hertzian and non-Hertzian methods and tangential contacts based on Kalker's linear theory and Polach theory.

Track irregularities tend to produce different magnitudes of force on the track. These forces on the track can result in three types of loads: (a) vertical, (b) lateral, and (c) longitudinal. Lateral loads are transverse to the track, while longitudinal loads are parallel to the track. Depending on their nature, loads can be (a) static loads, (b) quasi-static loads, and (c) dynamic loads. The dynamic loads may cause

- Irregularities in the track geometry
- Wear of the running surface
- Discontinuities on the running surface, which includes switches and frogs
- Dynamic forces, which appear in two categories: P1 and P2 forces. Frequencies of P1 forces range between 100 and 2000 Hz, are mainly impact forces.

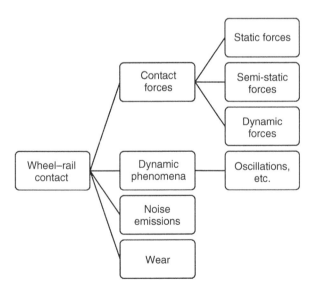

Figure 1.5 Wheel–rail contact impacts (Tzanakakis (2013). Reproduced with the permission of Springer)

P1 forces can cause, among other things, bolt hole failure and cracking of concrete ties and have minimal effects on the ballast or subgrade. P2 forces contribute to the degradation of track geometry and are classified in the frequency range between 30 and 100 Hz.

The contact between the wheel and the rail is the basic constitutive element of the railway dynamics (Table 1.4). For modeling purposes, two aspects are considered (Trzaska, 2012): (a) the geometric or kinematical relations of the wheel–rail contact and (b) the contact mechanical relations for the calculation of the contact forces. The wheel–rail contact provides insight into the formation of corrugation and other rail defects, like wear, crack growth, and others. The wear depends on tangential forces and creep age at the contact patch. Using mathematical analysis, it is possible to build a comprehensive and functional understanding of wheel–rail interaction, suspension and suspension component behavior, simulation, and experimental validation. This is beyond the scope of the current analysis.

Figure 1.6 shows various wheel–rail interfaces and their effects.

In wheel–rail contact there are three "zones" of contact, namely, Region A, Region B, and Region C, as shown in Figure 1.7. Region A is the contact between the central region of the rail crown and the wheel thread (conicity, hollow wear, and thermal loads), Region B is the extreme reference gage corner contact of the two-point contact, and, finally, Region C is the field side contact. At Region A,

Table 1.4 Vertical track forces.

Cause	Force	Symptom
Impact at rail welds	Rail: P1+P2	Rail fatigue failure
		Corrugations
		Pad degradation
		Tie cracking/movement
		Ballast degradation
		Weld fatigue
Vehicle/track interaction	Quasi-static	Track geometry deterioration
	Dynamic forces	Rail failure/fatigue
		Ballast failure/degradation
		Subgrade failure/degradation
Wheel irregularities	Wheel: P1+P2	Tie cracking
		Rail breaks
		Wheel cracks
		Ballast deterioration

Tzanakakis (2013). Reproduced with the permission of Springer.

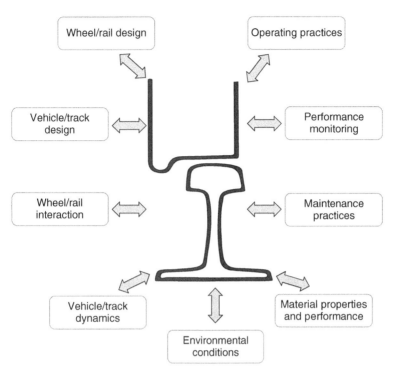

Figure 1.6 Wheel–rail interface

Figure 1.7 Regions of wheel/rail contact (Harris et al., 2001)

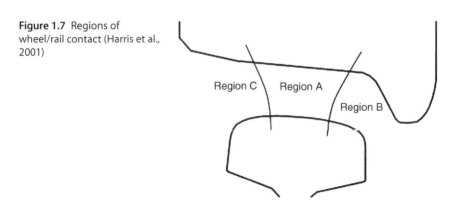

contact stresses are the lowest, lateral creepages and creep forces are low, and longitudinal creepages and forces are high, with reference to the wheel tread. In Region B, with reference to the wheel flange and gage corners, the contact between gage corner and flange fillet contact patch is small, with highest contact stresses, and there are three types of possible contact: (a) two-point contact, (b) single-point contact, and (c) conformal contact. In two-point contact, there is excessive creep relative to slip, high wear rates and material flaws, and less damage to the rail than the wheel. In the single-point contact, there are high longitudinal forces, the contact is most damaging to wheel and track, and head cracks are initialized and gage corners are broken out. During the conformal contact, gage corner and wheel flange wear to a common profile. In Region C, the field side contact, it is very difficult to optimize contact between wheel and rail due to the presence of high contact stresses.

Charles et al. (2008) developed a model-based condition monitoring applied to wheel–rail interaction parameter estimation, using the Kalman filter. The results appear to be very promising. The model was also capable of determining low adhesion detection and lateral track irregularities. Charles et al. (2008) in previous publications had used linear least square and other identification methods. Various techniques, including acoustic emission, have been used to monitor the continuous intensity of wheel–rail interaction. Also, various types of sensors are employed for this objective.

The current book will only attempt to use data science methods to address some of the output from wheel–rail experimental data. Thus, the approach will be more focused on probabilistic and statistical methods.

The derailment coefficient, the ratio between the lateral and vertical contact forces on the outside rail, has been shown to vary according to wheel–rail contact conditions, such as lubrication, time interval of train operation, rail temperature, and climate (Matsumoto et al., 2012). Currently, there are new systems that are used to collect data for that purpose. For example, data from wheel load sensors are sampled every 8 ms.

1.5.1 Switches and Crossings

Railway track switches and crossings (S&C) represent important components of the railway infrastructure. Generally, the S&C are referred to as turnouts, and their main function is to enable trains to change tracks. Train–track interaction at the turnouts is a major issue for both the design and maintenance of railway track systems (Bruni et al., 2009). Figure 1.8 shows the different types of S&C and Figure 1.9 shows the general structure of a crossing/switch. The nature of S&C makes them more expensive to maintain than regular straight and curved tracks (Zwanenburg, 2006):

- S&C have special components, like switch tongues, frogs, and slide plates, which are exposed to relatively high static and dynamic forces, making them experience high wear rates and specific deterioration.
- S&C have moving parts, which require extra regular inspections and maintenance actions.
- S&C form a potential safety hazard because of their moving parts and discontinuities that can create additional problems for failure, especially if they are not functioning well. This is a major safety issue.

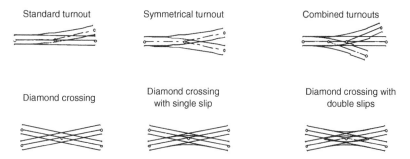

Figure 1.8 Different types of switches and crossings (Zwanenburg (2007). Reproduced with the permission of EPFL tous droits rèservès)

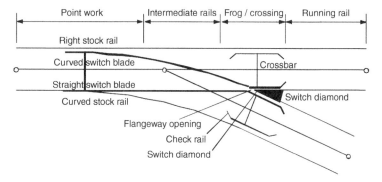

Figure 1.9 Schematized standard turnout and its components (Zwanenburg (2007). Reproduced with the permission of EPFL tous droits rèservès)

The wheel and rail profiles have a major influence on the deterioration of S&C. S&C degradation has an additional parameter compared with regular track parameters. Zwanenburg (2006) discussed the statistical approach, where analyses are based on loads, and directions of trains over specific S&C. Again, the author summarized specific factors that influence the degradation of the geometry of S&C. These are organized into three groups: (a) train parameters (e.g., axle load, speed, and number of axles); (b) track parameters (e.g., quality of the frog, quality of rail fasteners, tie type, ballast quality, and soil quality); (c) usage parameters (e.g., million gross tons (MGT), braking or acceleration, speed, and number of throws). Generally, based on the availability of data, different machine learning techniques can be used to model the deterioration of S&C.

1.6 Geometry Data

It is very important to distinguish between monitoring and inspection. Railway track inspection addresses safety concerns, while monitoring focuses on faults and supports efficient maintenance (Weston et al., 2015). Track geometry usually describes the position of each rail or track center line in space, in particular one rail with reference to the other. Therefore, the track geometry is a variation of lateral and vertical track positions in relation to the longitudinal plane or length parameter. The track consists of elements such as tangents, curves, transitions, S-shaped features, switches, and track irregularities (Haigermoser et al., 2015).

Track geometry consists of several parameters that have a major influence on ride quality and derailment risk. The Federal Railroad Administration (FRA) defines track geometry as the following geometry elements (FRA, 2014b):

- *Alignment*. Alignment is the projection of the track geometry of each rail or the track center line onto the horizontal plane. In other words, it is the relative position of the rails in its horizontal plane, and it is measured at the midpoint of a 62-ft chord. For tangent track, the alignment is equal to zero. For curved track the mid-chord offset of a 62 ft chord (in inches) is equal to the degree of curvature.
- *Crosslevel*. Crosslevel is the difference in elevation between the top surfaces of the two rails at any point of railroad track.
- *Gage*. Gage is the distance between to rail heads at right angles to the rails in a plane five-eighths of an inch below the top of the rail head. In the United States, the distance used is standard gage that is equal to 56.5 in for tracks containing up to 12° of curvature.
- *Profile*. Profile (also known as longitudinal leveling or surface) relates to the elevation along the longitudinal axis, which is an adherence to an established grade and the incidence of dips and humps. That is, it provides the elevation of the two rails along the track. The profile measurement is carried out in the midpoint of the 62-ft mid-chord.
- *Superelevation*. Superelevation (or cant) is the designed difference in the elevation between two rails in order to compensate for the effect of centrifugal forces in a curve. When that difference in the elevation does not fulfill the design requirements, it is called crosslevel.
- *Twist*. The difference in the crosslevel between two points of a fixed distance.
- *Warp*. The difference in the crosslevel between any two points within the specified chord length.

Track geometry is measured using automated systems and processed to give exceptions where measured parameters exceed a defined threshold. The current advances in the development of sensors have made track inspection systems that can compact and can be mounted underneath in-service vehicles possible. The large quantity of data collected by the sensors can be both advantageous and disadvantageous. Useful information can be obtained from the data, but at the same time there is the potential to lose critical information after the reduction of the data. Signal and processing techniques are used to derive important information from the output that can be used for maintenance planning and budgeting. In most cases, the track geometry defects should be within certain limits, which can be a function of maximum permissible speed. One example is the European Standard EN 13848-5, which provides limits for each type of defect. There are three main levels: (a) the immediate action limit (IAL) refers to the value that, if exceeded, requires imposing speed restrictions or immediate correction of the track geometry; (b) the intervention limit (IL) is the value that, if exceeded, requires corrective maintenance before the IAL is reached; and (c) the alert limit (AL) is the value that, if exceeded, requires that the track geometry condition be analyzed and considered in the regularly planned maintenance operations. The International Union of Railways (UIC) standard 518 (UIC 2005) designates three levels as follows: (a) QN1 quality level, which is the value that necessitates monitoring or taking maintenance actions as part of regularly planned maintenance operations; (b) QN2 quality level, which refers to the value that requires short-term maintenance action; and (c) QN3 quality level, which refers to the value above which it is no longer desirable to maintain the scheduled traffic speed. North American track safety standards (FRA, 2014a) present example of the geometry criteria.

The foot-by-foot track geometry condition is quantified using an index to represent the track geometry numerically, in order to reduce the data for practical analysis techniques. The provided overall qualities are referred to as the track quality indices (TQIs). The TQIs provide an aggregate level picture and do not identify individual defects very well. The standard deviation appears to be the most common track quality indicator, and in most cases it is calculated as the combination of various parameters. The following formula shows an example of a TQI:

$$\text{TQI} = \sqrt{w_{Al}\sigma_{Al}^2 + w_G\sigma_G^2 + w_{Cl}\sigma_{Cl}^2 + w_{T_p}\sigma_{T_p}^2},\tag{1.1}$$

where

Al: mean alignment
G: track gage
Cl: crosslevel
T_p: mean of the longitudinal level of both rails
w: weighting factor of the geometry parameters
σ: standard deviation of the individual parameters
The weighting factor is determined by the asset Management Manager.

Haigermoser et al. (2015) highlighted that the track quality is generally characterized by deviations of rail positions in three-dimensional space in terms of gage, crosslevel, longitudinal levels, alignment, and twist. Sometimes they are referred to as track irregularities.

Liu et al. (2015) presented the following definitions:

- *SD Index.* The SD index is associated with a track quality parameter and is calculated from measurement values for the parameter over a track segment (Equation 1.2). The larger the SD index is, the worse the track segment is in some aspect represented with the quality parameter.

$$\sigma_i = \sqrt{\frac{1}{n} \sum_{j=1}^{n} \left(x_{ij}^2 - \bar{x}_i^2 \right)}, \quad \bar{x}_i = \sum_{j=1}^{n} \frac{x_{ij}}{n}, \tag{1.2}$$

where
- σ_i: standard deviation of a quality parameter in millimeter
- x_{ij}: measurement value in millimeter for the parameter at the jth sampling point in the track segment
- n: number of sampling points in the track segment
- *Q Index.* ProRail of the Netherlands converts the SD index into a more universal form across different classes of tracks (Equation 1.3).

$$N = 10 * 0.675^{\sigma_i/\sigma_i^{80}}, \tag{1.3}$$

where
- N: Q index for a quality parameter over a 200 m long track segment
- σ_i: standard deviation for the quality parameter
- σ_i^{80}: the 80th percentile of standard deviations for 200 m long segments in a maintenance section with length ranging from 5 to 10 km.

The Q index ranges from 10 to 0. The larger the Q index, the better the track quality.
- *P Index.* The P index is adopted by Japanese railroads and is the ratio of the number of sampling points whose quality parameter measurements fall outside ±3 mm to the number of all sampling points in a track segment. It is applied to track segments of 100–500 m.
- *Track Roughness Index.* The track roughness index is used by Amtrak. In general, it presents the track roughness as the sum of squares of the amount of deviation measured over 20 m mid-chord offset (d^2), divided by the total number of measures (n) as shown in Equation 1.4.

$$R^2 = \sum_{i=1}^{n} \frac{d^2}{n} \tag{1.4}$$

- *Track Geometry Index.* The track geometry index TGI_i uses the measurement value space curve length L_i as shown in Equation 1.5.

$$TGI_i = \left(\frac{L_i}{L_0} - 1 \right) * 10^6 \quad L_i = \sum_{j=1}^{n-1} \sqrt{\left(x_{i(j+1)} - x_{ij} \right)^2 + \left(y_{j+1} - y_j \right)^2}, \tag{1.5}$$

where
- L_i: measurement value space curve length for a quality parameter over a track segment
- L_0: length of the track segment
- y_j: mile point of the jth sampling point on the track segment

- *CN's Track Quality Index.* The Canadian National Railway Company (CN) uses a second-order polynomial equation of the standard deviation σ_i of measurement values for a quality parameter over a track segment to assess its partial quality (Equation 1.6). The overall quality assessment is achieved by averaging six partial quality indices for gage, crosslevel, and left or right surface and alignment.

$$\text{TQI}_i = 1000 - C * \sigma_i^2, \tag{1.6}$$

- C: constant
- σ_i: standard deviation of measurement values

A larger track quality index implies the track segment has a better quality.

- *Overall Track Geometry Index (OTGI).* A variation of the TGI was developed by Sadeghi and Askarinejad (2010). Called the overall track geometry index (OTGI), it considers a normal distribution for each of the track geometry parameters (profile, alignment, gage, and twist). The authors combined individual geometry indices into a single expression presented in Equation 1.7.

$$\text{OTGI} = \frac{\frac{a}{2} \times \text{GI}^+ + \frac{a'}{2}\text{GI}^- + b \times \text{AI} + c \times \text{PI} + d \times \text{TI}}{\frac{a+a'}{2} + b + c + d}, \tag{1.7}$$

where
- GI^+ and GI^-: positive and negative gage indices, respectively
- AI: alignment index
- PI: profile index
- TI: twist index
- a, a', b, c, and d: model coefficients

- *Sweden TQI.* The Sweden National Railway assesses track geometry conditions as shown in Equation 1.8:

$$Q = 150 - 100 \left[\frac{\sigma_H}{\sigma_{Hlim}} + 2 \times \frac{\sigma_S}{\sigma_{Slim}} \right] \bigg/ 3, \tag{1.8}$$

where
- σ_H: standard deviation of the left and right profiles
- σ_S: standard deviation of the crosslevel, gage, and horizontal deviation
- σ_{Hlim}: standard deviation of the allowable σ_H
- σ_{Slim}: allowable value of σ_S

Table 1.5 shows indicators of different types of geometry defects

Table 1.5 Indicators for each type of defect according to EN 13848-5 (Teixeira and Andrade, 2014) Reproduced with the permission of Springer.

Type of defect	Indicators
Track gage	Nominal track gage to peak value
	Nominal track gage to mean track gage over 100 m
	Minimum value of mean value over 100 m (on tangent track and in curves of radius $R > 10,000$)
Longitudinal leveling	Mean to peak value (filtered in wavelength range 3–25 m)
	Mean to peak value (filtered in wavelength range 26–70 m) (only for train permissible speeds above 160 km/h)
	Standard deviation over a defined length, typically 200 m (filtered in wavelength range 3–25 m)
Horizontal alignment	Mean to peak value (filtered in wavelength range 3–25 m)
	Mean to peak value (filtered in wavelength range 26–70 m) (only for train permissible speeds above 160 km/h)
	Standard deviation over a defined length, typically 200 m (filtered in wavelength range 3–25 m)
Crosslevel	–
Twist	Zero to peak value (base length $l = 3$ m)

1.7 Track Geometry Degradation Models

Based on the existing literature, track geometry degradation models can be classified into two groups (Figure 1.10): deterministic and stochastic models. There are many contributions in this area in which different statistical techniques have been used in order to predict the track geometry degradation, which can be used as an input for determining the optimal schedule for maintenance activities. In this section, an overview of the contributions in literature regarding these track degradation models is presented, highlighting the main characteristics of each model and data collection technologies as well as discussing the findings and trends regarding track degradation models.

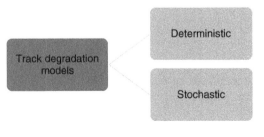

Figure 1.10 Classification of track geometry models based on parameters' uncertainty

1.7.1 Deterministic Models

In general, a deterministic model assumes that both the input and the output of a system are constant values, so there is no uncertainty involved. That means that the output of the model only depends on the initial condition of the system and the parameter values. In track geometry degradation, those deterministic models can be linear or nonlinear (polynomial, exponential, etc.), and they are created with the assumption that model parameters are fixed values. There are literature reviews that address deterministic models for track geometry degradation, in which the works of Oberg (2006), Guler (2013), and Dahlberg (2001) are highlighted. In this section, a review of the contributions of track geometry degradation models is presented.

1.7.1.1 Linear Models

In terms of linear track geometry degradation models, it is assumed that the model is linear in its parameters and that track degrades in a constant rate usually referred to in terms of in./MGT. Once it reaches a specified intervention threshold, maintenance activities are performed, such as tamping in order to achieve the desired roughness level (Figure 1.11). Usually, the intervention thresholds are determined based on railroads. In the United States, the FRA establishes the tolerances in terms of track safety requirements.

There exist in the literature several contributions of a linear representation of track geometry degradation, as presented below.

Chang et al. (2010) incorporated multilinear components in the track degradation model. The authors highlight the characteristics of the track geometry degradation in terms of three elements: (a) periodicity, which means that track geometry defects change patterns over the same track section and are the same between two adjacent track maintenance; (b) multistage, which means

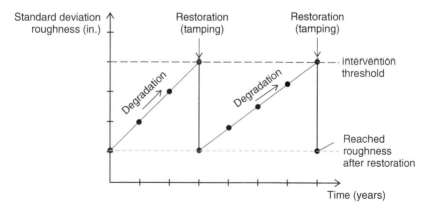

Figure 1.11 Linear representation of track geometry degradation and restoration based on the standard deviation of roughness

that the track geometry degradation rate varies from beginning to the end; and (c) experimental, which refers to the fact that track geometry degradation rate increases as passing tonnage increases. The multilinear model is presented in Equation 1.9 as follows:

$$\sigma_{TLD} = a_n + b_n T, \tag{1.9}$$

where

- σ_{TLD}: standard deviation of longitudinal leveling defects
- b_n: slope of line n
- a_n: intercept of line n
- T: cumulative passing tonnage from last maintenance to the present day

1.7.1.2 Nonlinear Models

Nonlinear models, on the other hand, assume that the degradation model is not linear in its parameters, so, in counterpart to linear models, the track roughness can be either a polynomial or exponential function, among others (Figure 1.12).

1.7.2 Stochastic Models

In general, a stochastic model assumes uncertainty in the model analysis.

Andrade and Teixeira (2012) implemented Markov chain Monte Carlo (MCMC) to estimate track geometry deterioration model parameters. The objective is to evaluate the uncertainty behavior of the infrastructure through its life cycle. The data was taken from the railway Lisbon–Porto in Portugal. In terms of modeling purposes, the authors considered the standard deviation of the longitudinal leveling defects. The track geometry deterioration model

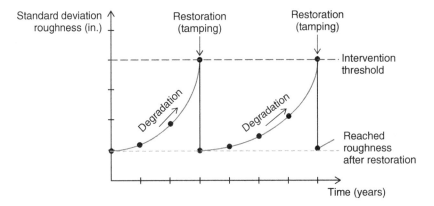

Figure 1.12 Nonlinear representation of track geometry degradation and restoration based on the standard deviation of roughness

follows the linear relationship using 200 m of track section, as shown in Equation 1.10.

$$\sigma_{LD} = c_1 + c_0 T, \tag{1.10}$$

where

- σ_{LD}: standard deviation of the longitudinal leveling defects (mm)
- c_1: initial standard deviation of the longitudinal leveling defects right after upgrade actions (mm)
- c_0: deterioration rate (mm/100 MGT)
- T: cumulative tonnage after track upgrade (100 MGT)

In terms of the Bayesian analysis, the authors assumed informative prior distribution by using elicitation. For the prior estimation of parameters c_1 and c_0 at the design stage, the authors used inspection data from the railroad Lisbon–Porto. By using a hypothesis test, the deterioration rate c_0 followed a log-normal distribution, and the initial standard deviation of the longitudinal leveling defects c_1 followed a bivariate log-normal distribution. In terms of the MCMC, the simulations were run considering the track sections divided into four groups: plain track, bridges, stations, and switches. The MCMC convergence was controlled by using a burn-in period length of 2000 values. The thinning consisted of 10,000 values with a lag $L = 500$; it was made in order to eliminate the autocorrelation of the time series.

Andrade and Teixeira (2013) considered a hierarchical Bayesian approach where they assumed that the standard deviation of longitudinal level defects was normally distributed and where the mean was composed of the following elements: (a) the constant linear evolution with MGT, (b) the initial standard deviation of longitudinal defects, (c) the disturbance effect of the initial standard deviation of longitudinal level defects after each tamping operation, and (d) the differentiate of renewed track and nonrenewed track sections. In addition, the authors assumed non-informative priors using inverse gamma distributions.

Likewise, the work of Andrade and Teixeira (2011) defined track geometry defects in terms of the elements described above, as well as the linear representation of the standard deviation of the longitudinal leveling defects. For parameter estimation, the authors used the Kolmogorov–Smirnov goodness-of-fit test. In addition, a Monte Carlo simulation was performed in order to assess the uncertainty regarding the tamping cycle for each track section group and specific quality level, concluding that ballasted deck bridges and switches require far more frequent tamping actions compared with stations and plain track.

Audley and Andrews (2013) also present a linear degradation model as the relationship presented in Equation 1.11. For parameter estimation, the authors utilized a maximum likelihood estimation method.

$$\sigma = A + Bt, \tag{1.11}$$

where

- σ: vertical alignment (mm)
- A: intercept (mm)
- B: deterioration rate (mm/day)

In terms of Markov models, Yousefikia et al. (2014) presented a Markov model in order to assess tram track condition and predict maintenance actions for Melbourne tram track data. For that purpose, the authors defined seven states for track condition named as follows:

- Normal
- Maintenance limit – degraded failure undetected
- Maintenance limit – degraded failure inspected
- Action limit – degraded failure undetected
- Action limit – degraded failure inspected
- Safety limit
- Repaired

On the other hand, He et al. (2013) considered a stochastic track geometry degradation model, considering factors such as the monthly traffic MGT traveling through track lot k ($X_{1t}(t)$), the monthly total number of cars ($X_{2t}(t)$), the monthly total number of trains ($X_{3t}(t)$), and the number of inspection runs in sequence since the last observed red tag geometry defect ($X_{4t}(t)$). Equation 1.12 shows the representation of the model.

$$\log\left(\frac{Y_k(t + \Delta t) - Y_k(t)}{\Delta t Y_k(t)}\right) = \alpha_0 + \alpha_1 X_{1k}(t) + \cdots + \alpha_p X_{pk}(t) + \varepsilon_k(t),$$

$$\forall k = 1, \cdots, N \tag{1.12}$$

where

- $Y_k(t)$: aggregated geometry defect amplitude of the track lot k at inspection time t
- N: total number of track lots
- $X_{pk}(t)$: pth external factor or predictor for kth track lot at inspection time t
- α_0: model intercept
- α_i: coefficients for ith X factor. $i = 1, \ldots, p$
- $\varepsilon_k(t)$: random error. $\varepsilon_k(t) \sim N(0, \sigma^2)$

In addition, Vale and Lurdes (2013) performed a stochastic model based on Dagum distribution, which is used for describing the track geometry degradation process over time. Dagum distribution is a function of the input data and the model parameters. For parameter estimation, the authors used the maximum likelihood method.

Meier-Hirmer et al. (2006) presented a gamma process for track degradation and a classification method based on regression trees using environmental variables, such as type and height of ballast, maximum speed, weather conditions, type of rail, and accumulated tonnage since ballast renewal, among others. In the paper, the authors considered longitudinal leveling defects as the track failure mechanism for track geometry defects.

Oyama and Miwa (2006) developed a model for measuring track surface irregularities and predicting track maintenance operation effects based on a logistic distribution. The transition process of surface irregularities model is composed of two processes: degradation and restoration. For the degradation model, the exponential smoothing method was used to predict the increasing trend of parameter b (standard deviation of logistic distribution) and expresses the typical characteristics of the track surface irregularities. For the restoration model, the β values before and after tamping operations were compared.

Table 1.6 attempts to summarize various models and the authors based on the work presented by Galvan-Nunez (2017).

1.7.3 Discussion

Based on the literature review (Table 1.6), the following conclusions are made:

- Most deterministic models focus on the study of differential ballast settlement as the main track geometry failure mechanism.
- There is no uniqueness in the used terminology with regard to track geometry degradation.
- Although the research trend suggests to consider uncertainty in track geometry degradation models, there is still no consensus in literature about such models. This is verified with the wide number of publications in the area that use different track geometry parameters. Based on that, it can be seen that the study of track geometry degradation is an open field that needs to be improved.

1.8 Rail Defect Data

Rail defects are almost always caused by fatigue from wheel–rail interactions and the presence of defective materials. Cannon et al. (2003) noted that fatigue failure in rail develops in three basic phases: first a fatigue crack initiates, and then it grows in size, and finally in the absence of control or maintenance, the rail breaks. The authors presented three broad groups of rail failure as follows:

- Those originating from rail manufacture – for example, tache or kidney defect
- Inappropriate handling, installation, and use – for example, wheel burn defects
- Those caused by exhaustion of rail steel's inherent resistance to fatigue damage – for example, head checking, squats, and others.

Table 1.6 Summary of literature review[a].

Author	Degradation model	Model parameters
Iyengar and Jaiswal (1995)	Random field	Absolute vertical profile, unevenness data
Meier-Hirmer et al. (2006)	Gamma process	NL: longitudinal leveling, α: Gamma parameter (constant), β: Gamma parameter (constant)
Oyama and Miwa (2006)	Exponential smoothing	b: standard deviation of logistic distribution/track surface irregularities
Veit (2007)		
Chang et al. (2010)	Multi-stage linear regression	σ_{TLD}: standard deviation of longitudinal level irregularity, b_n: slope of line n, a_n: intercept of line n, T: cumulative passing tonnage from last maintenance to the present day
Xu et al. (2011)	Linear regression	
Berawi et al. (2010)	Linear regression	
Quiroga and Schnieder (2012)	Exponential function	NL: longitudinal leveling, NL_{init}: initial longitudinal leveling, t: time, t_n: time at which the last tamping operation took place, b_n: log-normally distributed variable, $en(t)$: measurement noise/normally distributed variable, n: number of tamping interventions
Andrade and Teixeira (2011)	Linear regression	c_1: Initial standard deviation after renewal or tamping operations, c_0: Deterioration rate (mm/100 MGT), T: Cumulative tonnage between tamping operations (100 MGT)
Andrade and Teixeira (2012)	Linear regression	c_1: initial standard deviation after renewal or tamping operations, c_0: Deterioration rate (mm/100 MGT), T: cumulative tonnage between tamping operations (100 MGT)

Table 1.6 (Continued)

Author	Degradation model	Model parameters
Andrade and Teixeira (2013)	Hierarchical Bayesian	Y_{svkl}: standard deviation of longitudinal level defects, T_{svkl}: accumulated tonnage since last tamping or renewal operations, b_{svk}: deterioration rate (constant linear), a_{svk}: initial standard deviation of longitudinal level defects, d_{sv}: initial standard deviation of longitudinal level defects after each tamping operations, N_{svkl}: number of tamping operations performed since last renewal
Vale and Lurdes	Dagum model	Standard deviation longitudinal level, Dagum distribution shape and scale parameters
Audley and Andrews (2013)	Linear regression	
Yousefikia et al. (2014)	Markov chain	Markov chain states: normal, maintenance limit. Degraded failure undetected, maintenance limit. Degraded failure inspected, action limit. Degraded failure undetected, action limit. Degraded failure inspected, safety limit. Repaired
Guler (2013)	Artificial neural networks	−Gradient (%) curvature (1/R) (1/m), crosslevel (mm), speed (km/h), age (years), rail type (kg/m), rail length (m), tie type

a) Galvan-Nunez (2017).

Rail defect management is one of the most important tasks in ensuring the reliable and safe operation of rail transport. Two major objectives of rail defect management are to (a) detect and rectify rail defects before they cause rail breaks and to (b) reduce and eliminate rail defects. Now with large amounts of data available from different railway agencies, the use of traditional statistical methods to address these objectives may not be adequate. The defect management systems include reporting and classifying rail failures, inspection and actions, rail failure statistics, and other important information. The FRA (FRA, 2015) presented an extensive catalog of different rail defects and their nomenclatures. Broken rails occurring when traffic is at track speed can be costly present other safety issues (Zarembski and Palese, 2005).

Figures 1.13 and 1.14 show cross sections of rail and different planes. Tables 1.7–1.13 show different rail defects.

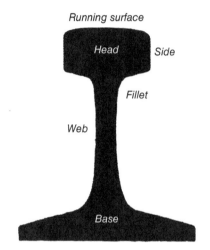

Figure 1.13 Cross-section of a rail

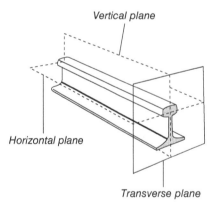

Figure 1.14 Transverse, vertical, and horizontal places of track

Table 1.7 Rail defects classification.

Defect name	Characteristics
Plastic deformation	Will always be present in rail–wheel contact On a microscopic scale – close to the rail surface On a macroscopic scale – change of profile shape Work hardening and adaption to loading condition
Wear	Interaction between two surfaces results in dimensional loss of one solid Continuous material removal from the rail surface due to interaction between wheel and rail Several empirical wear laws do exist Types of wear (in the wheel–rail contact): adhesive wear, abrasive wear, fatigue wear, corrosive wear
Corrugation	Wave structure on the rail surface (tangent/curve) Short wave (25–80 mm wavelength) or long wave (100–300 mm) corrugation More detailed classification possible Combination of wear and plastic deformation Damage of other track components possible
Head checks/gage corner cracking	Periodic cracks at the gage corner Sometimes also cause periodic cracks on the running surface Grade-dependent crack spacing Can cause detail fracture if not treated
Flaking and spalling	Head checks can combine, causing material to break out of the rail surface Head checks – flaking – spalling
Shelling	Originates underneath the rail surface Delamination of rail material – crack will surface at gage corner and cause material to break out High loading conditions favor formation
Squats	Widening of running band Typical kidney-like shape and V-crack Three different classes (light, medium, severe) Difficult to detect in early stages by automated means Singular event or epidemic occurrence Can cause detail fracture Mixed traffic to low load conditions (low wear)
Detail analysis – severe squat	V-crack with opposite surfacing crack on running band Strong widening of running band and extended longitudinal extension Point of crack origin at tip of V Bowl-like growth of subsurface crack Break outs on rail surface
Wheel burn	Occurs in pairs directly opposite to each other (both rails) Continuous slipping of locomotive wheel set High temperature input to rail surface Formation of hard and brittle martensite layers (thick layers) Massive damage to rail surface

Table 1.8 Transverse defects.

Defect type	Detection	Comments
Transverse fissure	Can be detected by 70° transducer beams	Growth is initially slow, until defect 20–25% Failure almost occurs before defect becomes visible
Compound fissure	Can be detected by 70° transducer beams	Growth is normally slow until the defect reaches 30–35% Failure occurs before defect is visible Complete break of the rail across head, web, and base
Detail fracture Shelling Head check	Can be detected by 70° transducer beams	Growth is slow until defect reaches 15% Failure occurs before defect is visible
Engine burn fracture	Can be detected by 70° transducer beams	Growth is normally slow until the defect reaches 10–15% Failure occurs before defect is visible
Welded burn fracture	Can be detected by 70° transducer beams	Growth is normally slow until the defect reaches 15–20% Failure occurs before defect is visible

Table 1.9 Longitudinal defects.

Defect type	Detection	Comments
Horizontal split head	0° and 45° transducer beams	May develop into a compound fissure
Vertical split head	0° and 45° transducer beams	Initially not visible Widening of the head for the length of the split
Shear break	0° and 45° transducer beams	Growth is sudden Initially not visible

Table 1.10 Web defects.

Defect type	Detection	Comments
Head and web separation	0° and 45° transducer beams	Growth is rapid Entire length of the rail is usually weakened
Split web	0° and 45° transducer beams	Growth is rapid Entire length of the rail is usually weakened
Piped rail	0° and 45° transducer beams	Growth is usually slow Heavy loads may accelerate growth Rail is weakened for the distance of the pipe
Web and head separation	0° and 45° transducer beams	Growth usually occurs in gradual stages Rail is weakened for a distance in excess of the progressive separation

Table 1.11 Base defects.

Defect type	Detection	Comments
Broken base	0° and 45° transducer beams	Weakened section
Base fractures	0° and 45° transducer beams	Growth relatively slow until the defect progresses from the edge of the base into the rail

Table 1.12 Surface defects.

Defect type	Detection	Comments

Corrugation
  Short pitch
  Long pitch
	Visual inspection	Not a serious hazard
		Short pitch between 30 and 90 mm wavelength
		Long pitch between 150 and 200 mm wavelength
		Can be rectified by grinding

Rolling contact fatigue	Visual inspection	Cracks initiate close to rail surface Cracks can spread across the rail head Can be rectified by grinding
Shelling	Visual inspection	Occurs frequently in curve territory Transverse representation may occur

Squats
  Running surface
  Gage corner
	Visual inspection	Type of rolling contact fatigue
		Cracking when in moderate to severe stages
		Can develop in track with either timber or concrete ties
		Often develops in switches and turnouts
		Rail replacement needed in severe cases

Table 1.13 Wheel burns.

Defect type	Detection	Comments
Wheel burns	Visual inspection	Does not grow Damage may cause roughening of traffic Can cause transverse separation May develop into thermal cracks

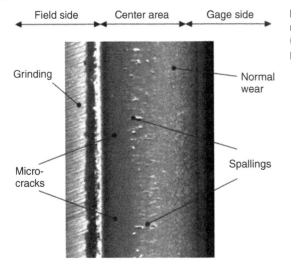

Figure 1.15 Surface regions of rail head (Huber-Mörk et al. (2010). Reproduced with the permission of Springer)

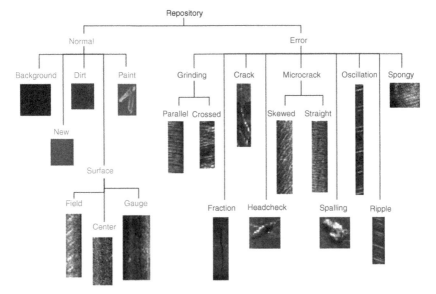

Figure 1.16 Repository of rail head surface classes: normal or noncritical surface (Huber-Mörk et al. (2010). Reproduced with the permission of Springer)

Figure 1.15 shows the surface regions of the rail head. Figure 1.16 shows different surface classes.

Figure 1.17 shows an example of defects per mile for track section AB. The defect rate is nonlinear. Figure 1.18 shows the percentage of all defects per mile.

Rail track defects per mile or defect initiation grows with age but the growth rate is not linear. The crack also grows nonlinearly from initiation to failure. Figure 1.17 is a section of rail track.

The Weibull distribution has been the method used in the probabilistic analysis of rail defect occurrences within the rail track. It has been shown that

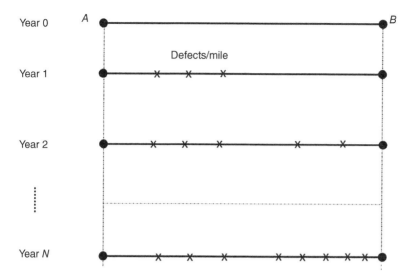

Figure 1.17 Rail defects per mile

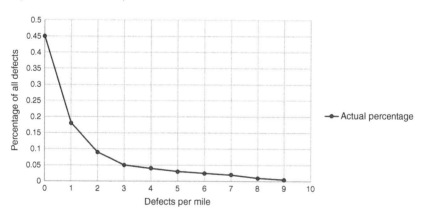

Figure 1.18 Rail defects distribution

the past defect history appears to be important for the prediction of future rail defects (Palese and Wright, 2000). Figure 1.18 demonstrates statistical rail defect distribution.

1.9 Inspection and Detection Systems

There are various methods used to detect defects; these include the following:

- Ultrasonic testing from non-destructive test vehicle.
- Visual inspection methods by the track maintenance team. They also involve the use of ultrasonic equipment.
- Dye penetration and magnetic particle method. This is usually used for surface defects.

- Eddy current testing. This is a noncontact method of identifying surface-breaking or near-breaking defects.
- Radiography. This is used for the examination of aluminothermic welds that contain irregular, nonplanar defects and defects that are very difficult to detect using ultrasonic methods.

There are other new technologies, which include the following:

- Low frequency eddy current sensors to locate deeply buried defects. This type of detection makes the application of deep neural networks a viable technique in analyzing the data.
- Laser generation and reception of ultrasonic waves to enable noncontacting inspection (Allipi et al., 2000).
- Longitudinal guided waves, which allow locomotives to scan the track ahead.
- High-speed ultrasonic testing.
- Automated vision systems – their application is restricted to surface defects.
- Machine vision and profilometer technology.
- Ultrasonic rail flaw testing (Sawadisavi, 2010).
- Radiographic inspection or X-ray diffraction measurements for rolling contact fatigue layer (Matsui, 2014).
- Ground-penetrating radar (GPR) applications. The GPR inspection is focused on the thickness of rail track ballast, subsoil materials penetrating the ballast, and properties of the subgrade materials.

Currently there are also hybrid systems, combinations of human inspection and automated inspection devices. The general rationale is that human inspectors can (a) perform more detailed inspection but not necessarily objective, (b) provide better decision-making on the found defect, and (c) search for a larger variety of defects. The automated inspection, on the other hand, can (a) operate at a higher speed, (b) better locate small variations from the normal pattern, (c) limit human errors and bias, and (d) be very objective, consistent, and precise.

Carr (2014) discussed the application of autonomous track inspection technology that inspects the track from revenue service trains using unattended instrumentation. For example, the autonomous track geometry measurement systems (ATGMS) are mounted on the rail car and have remote access. They provide efficient inspections at a much lower cost and can identify changes in patterns very early through frequent inspections. Carr (2014) presented various inspection systems that are in operation and those that will be in operation soon. Table 1.14 shows the various inspection systems. There are systems available that can automatically detect visible anomalies on the surface of the rail, a missing fastener element, or a cracked or damaged tie and measure the ballast profile and hence the volume of the ballast. There are also other technologies, like the acoustic bearing detector (ABD) that uses sound

Table 1.14 NDT techniques for the rail industry.

NDT technique	Systems available	Defects detected	Performance
Ultrasonics	Manual and high-speed systems (up to 70 km/h)	Surface defects, rail head internal defects, rail web, and foot defects	Reliable manual inspection but can miss rail foot defects. At high speed can miss surface defects smaller <4 mm as well as internal defects, particularly at the rail foot
Magnetic flux leakage (MFL)	High-speed systems (up to 35 km/h)	Surface defects and near-surface internal rail head defects	Reliable in detecting surface defects and shallow internal rail head defects but cannot detect cracks smaller than <4 mm. MFL performance deteriorates at higher speeds
Pulsed eddy current (including field gradient imaging)	Manual and high-speed systems (up to 70 km/h)	Surface and near-surface internal defects	Reliable in detecting surface-breaking defects. Adversely affected by grinding marks and lift-off variations
Automated Visual Inspection	Manual and high-speed systems (up to 320 km/h)	Surface-breaking defects, rail head profile, corrugation, missing parts, defective ballast	Reliable in detecting corrugation, rail head profile missing parts, and defective ballast at high speeds. Cannot reliably detect surface-breaking defects at speeds >4 km/h. Cannot assess the rail for internal defects
Radiography	Manual systems for static tests	Welds and known defects	Reliable in detecting internal defects in welds difficult to inspect by other means. Can miss certain transverse defects
Electromagnetic acoustic transducers	Low-speed hi-rail vehicle (<10 km/h)	Surface defects, rail head, web, and foot internal defects	Reliable for surface and internal defects. Can miss rail foot defects. Adversely affected by lift-off variations

(Continued)

Table 1.14 (Continued)

NDT technique	Systems available	Defects detected	Performance
Long-range ultrasonics	Manual systems and low-speed hi-rail vehicle systems (<10 km/h)	Surface defects, rail head internal defects, rail web, and foot defects	Reliable in detecting large transverse defects (>5% of the overall cross-section)
Laser ultrasonics	Manual and low-speed hi-rail vehicle systems (<15 km/h)	Rail head, web and foot defects	Reliable in detecting internal defects. Can be affected by lift-off variations of the sensors, difficult to deploy at high speeds
Alternating current field measurement	Manual systems (hi-speed system under development)	Surface-breaking defects	Reliable in detecting and quantifying surface-breaking defects. Cannot detect subsurface defects. Very good tolerance to lift-off variations
Multifrequency eddy current sensors	Manual system. Static and slow speed.	Surface and near-surface defects.	Limited experiments conducted. Has the potential to reliably quantify defects detected
MAPS	Manual system. Static and walking speed tests	Residual stresses	Results comparable to X-ray diffraction. Commercially available

(Papaelias et al., 2008)

properties generated by specific component defects, the hot box detector (HBD) that evaluates the bearing temperature history and other defect issues based on temperature outliers, and the cracked wheel/axle detector that consists of rail-mounted sensors detect different tones generated by normal versus flawed wheel axles (Rose, 2015).

The Alliance for Innovation and Infrastructure (AII, 2015) noted the following technologies for improving track integrity:

- *Track Integrity Sensors.* These can be used in both the rail and the ballast. It broadcasts anomalies to monitoring stations, and it is useful for maintenance in demand.
- *Ballast Integrity Sensors.* These provide continuous, real-time monitoring of subgrade movement in reference to the track structure.
- *Autonomous Track Geometry Measurements.* These measure and record track geometry remotely from an autonomous rail car in a regular train service.
- *Gage Restraint Measurement Systems (GRMS).* These are systems that measure rail motion under a combined vertical and lateral load to detect weak ties and fasteners. Their use allows inspectors to identify specific conditions at a specific location of tracks.
- *Ultrasonic and Induction Rail Testing.* Ultrasonic uses waves sent in angles that are reflected back to transducers and analyzed.

1.10   Rail Grinding

Zarembski and Palese (2005) presented a comprehensive overview and detailed technical description of rail grinding. The authors defined railway grinding as the removal of small amounts of metal from the top of the rail through the use of abrasive grinding materials with the rotating properties of the grinding motors. Rail grinding is an integral part of railway track routine maintenance. Some reasons for rail grinding are controlling (a) surface fatigue, (b) weld dipping, (c) plastic flow, (d) rail wear, (e) wheel wear and fatigue, (f) truck hunting, and (g) wheel/rail noise and (h) improving reliability of ultrasonic rail test. The success of the grinding operations depends on the characteristics and condition of the abrasive wheel, the applied pressure, and the speed of and angle between the grinding stones and the rail (Cuervo et al., 2015; Hartsough et al., 2016). There are four principal types of rail grinding (Elaina, 2015):

- *Corrective or Defect Grinding.* This is primarily the removal of the rail defects that have developed, which include (a) corrugations, (b) gross plastic flow, and (c) rolling contact fatigue defects. This involves aggressive grinding procedures that aim to remove a considerable amount of metal between 0.5 and 6 mm at long intervals determined by the severity and cluster.

- *Transitional Grinding.* This is less intensive than corrective grinding. The main objective is to provide a partial corrective grind to prepare the tracks for a preventive grind.
- *Preventative or Cyclic Grinding.* The objective of this type of grind is to eliminate or at least control rail defects and maintain the surface condition and preferred rail profiles. It usually involves removing about a minimum of 0.2 mm at more frequent and controlled intervals. The frequency is often determined by estimation and experience, so there is a fair amount of subjectivity. Preventative grinding can prolong rail life.
- *Special Grinding.* This involves the use of grinding to achieve a specific objective, for example, establishing the rail profiles to reduce the rate of wheel hollowing or to provide a very smooth rail contact surface to reduce noise and other dynamic effects generated by wheel–rail contact.

There are quite a number of optimization issues encountered in preventive maintenance metal removal because the optimal metal removal rate is dependent on different factors, including (a) tonnage since the last grinding cycle, (b) axle load, (c) traffic type, (d) rail metallurgy, (e) track grade, (f) track superelevation, (g) track curvature, (h) track gage, (i) track support structure, (j) friction management, and (k) environmental factors.

Bremsteller (2014) depicted critical defect depth and MGT. It was concluded that surface defects do not develop linearly and that doubling MGT results can result in triple damage depth. Figure 1.19 shows the graph of critical depth

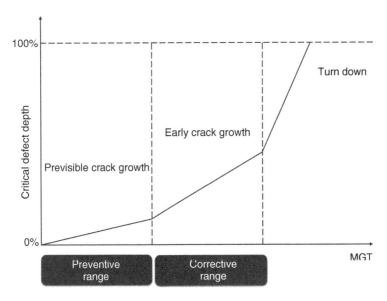

Figure 1.19 Rolling contact fatigue. Courtesy: Johannes Bremsteller

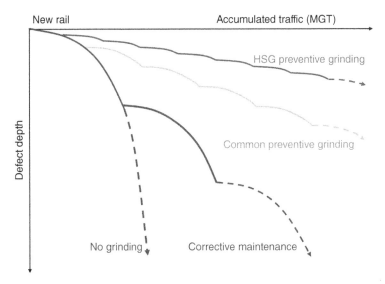

Figure 1.20 Different rail maintenance strategies. Courtesy: Johannes Bremsteller

and MGT. Again the authors presented a graph comparing different rail maintenance strategies and defect depth (Figure 1.20). One major analytical issue based on the frequency curves is how effective the grinding process is at eliminating certain types of defects.

Palese et al. (2004) developed the grinding quality index (GQI) to evaluate the effectiveness of rail profile grinding and to check the difference between ground rail profile and the desired profile. By calculating the GQI before and after grinding, the effectiveness of the grinding operation can be assessed. The accuracy of GQI depends on the following:

a) Method of profile captured
b) Quality of rail image
c) Calibration process
d) Train speed during measurement
e) Environmental conditions and factors

Mathematically, the grinding process can be expressed as follows:

$$y_{p_i} = p_i(x),\qquad(1.13)$$

$$y_{R_i} = r_i(x),\qquad(1.14)$$

where y_{p_i} is the measure profile, $y_{R_i}, y_p \geq y_R, \forall x$.

The area between profile and template can be expressed as follows:

$$\text{Area} = [P_i(x) - r_i(x)]\Delta x\qquad(1.15)$$

The GQI

$$G_i = 100 \frac{A_b}{A_a + A_b},$$ (1.16)

where

- A_a: Area difference profile above acceptable envelope
- A_b: Area difference profile below acceptable envelope
- G_i is between 0 and 100

1.11 Traditional Data Analysis Techniques

Various data analytical techniques have been used to analyze the output of rank inspection systems. Table 1.15 shows some of the techniques that have been used. Some cases are hybrid methods that involve the combination of different analytical techniques, like traditional neural network and wavelets application. Other hybrid applications include edge detection methods and the support vector machine (SVM). Bhaduri (2013) used a wavelength intensity algorithm as an extraction algorithm and used the SVM as a classifier. Arivazhagan et al. (2015) discussed various automatic crack detection techniques based on morphological operations, fractal analysis, and edge detection schemes.

Table 1.15 Automated visual railway component inspection methods.

Components	Defects	Features	Decision methods
Fasteners	Missing	DWT	Three-layer NN
Fasteners	Missing	Edge density	Threshold
Fasteners	Broken	DWT	Threshold
Joint bars	Cracks	Edges	SVM
Fasteners	Missing/defective	Intensity	OT-MACH corr.
Fasteners	Broken	Haar	Adaboost
Fasteners	Missing	Direction field	Correlation
Ties/turnouts	–	Gabor	SVM
Tie plates	Missing spikes	Lines/haar	Adaboost
Fasteners	Missing/defective	Haar	PGM
Concrete ties	Cracks	DST	Iterative shrinkage
Fasteners	Defective	Harris-Stephen, Shi-Tomasi	Matching errors
Fasteners	Missing/defective	HOG	SVM
Concrete ties	Tie condition	Intensity	Deep CNN

Gibert et al. (2015). Reproduced with the permission of IEEE.

Gibert et al. (2015) discussed how multitask learning, an inductive transfer learning method in which two or more machine learning techniques are trained, can be applicable in track output data analysis. The main idea of this approach is that each task can benefit from reusing knowledge that has been learned while training for other tasks.

GPR data analysis involves analysis of images and signal processing tools to address the information obtained. Novel multivariate signal and image processing techniques are currently used to automatically detect and quantify various anomalies in the ballast and other subsurface conditions. Li et al. [Evaluation of GPR Technologies for Assessing Track Substructure Conditions] presented the results of GPR railway track applications. Li et al., attempted to tie the geometrical characteristics of the track with the GPR results. Ibrekk (2015) used GPR to detect subsurface track body anomalies like ballast pockets and animal burrows and to map the distribution of water in the track body. Also, Ibrekk (2015) discussed some GPR studies and railway applications: (a) measuring ballast layer thickness; (b) determining the degree of ballast fouling; (c) locating hidden objects; (d) detecting ballast anomalies that include subgrade penetration, ballast pockets, and mud pumping; and (e) detecting frost susceptibility and ice lens formations. In GPR analysis, sources of unwanted signal noise, or any abnormal signal, should be identified and removed from the output before initiating any analysis. The short-time Fourier transform has been used in ballast fouling surveys. SVM and k-neural network were also used in GPR ballast applications (Shao et al., 2011).

Recently, Einbinder (2015) used multivariate adaptive regression splines (MARS) to develop models that connect rail defects in terms of MGT to geometry defect variables. One major characteristic of the models developed by Einbinder (2015) is that the models were able to develop a set of equations between specific MGT ranges.

1.11.1 Emerging Data Analysis

Table 1.16 compares big data and traditional data. Most of the current analysis falls under traditional data analysis. As data becomes extremely large, different properties of the data will change, and hence one needs to devise new methods, like ABC algorithms, to handle and solve problems emerging from railway track monitoring and control output. Critical questions need to be addressed including how one can use the MapReduce architecture to solve problems.

1.12 Remarks

Most of the current data analysis techniques may not be appropriate in cases where there are large amounts of data present. It is also not clear if the current

Table 1.16 Big data versus traditional data.

	Traditional data	Big data
Volume	GB	Constantly updated (TB or PB currently)
Generated rate	Per hour, day, …	More rapid
Structure	Structured	Semi-structures or unstructured
Data source	Centralized	Fully distributed
Data integration	Easy	Difficult
Data store	RDBMS	HDFS, NoSQL
Access	Interactive	Batch or near real time

techniques will be efficient in solving problems on streaming data. Therefore, there is a need to develop a framework to analyze track data. As noted from the previous paragraphs, all the data, geometry or rail defect data, have most of the following characteristics:

- Massive data sets
- Unstructured data and heterogeneous databases
- Information in the form of images
- Poor quality of data
- Multiresolution and multisensor data
- Spatiotemporal data
- Noisy data
- Missing data
- Streaming data

References

AII. Back on track: bringing rail safety to the 21st century. Technical report, Alliance for Innovation and Infrastructure, 2015.

C. Alippi, E. Casagrande, F. Scotti, and V. Piuri. Composite real-time image processing for railways track profile measurement. *IEEE Transactions on Instrumentation and Measurement*, **49**(3):559–564, 2000.

A. R. Andrade and P. F. Teixeira. Uncertainty in rail-track geometry degradation: Lisbon-Oporto line case study. *Journal of Transportation Engineering ASCE*, **687**:193–200, 2011. doi: 10.1061/(ASCE)TE.1943-5436.0000206.

A. R. Andrade and P. F. Teixeira. A Bayesian model to assess rail track geometry degradation through its life-cycle. *Research in Transportation Economics*, **36**(1):1–8, 2012. doi: 10.1016/j.retrec.2012.03.011.

A. R. Andrade and P. F. Teixeira. Hierarchical Bayesian modelling of rail track geometry degradation. *Proceedings of the Institution of Mechanical Engineers, Part F: Journal of Rail and Rapid Transit*, **227**(4):364–375, 2013. doi: 10.1177/0954409713486619.

S. Arivazhagan, R. N. Shebiah, J. S. Magdalene, and G. Sushmitha. Railway track derailment inspection system using segmentation based fractal texture analysis. *Journal on Image and Video Processing*, **6**(1):1060–1065, 2015. http://ictactjournals.in/paper/IJIVP_Paper_2_pp_1060_1065.pdf.

M. Audley and J. Andrews. The effects of tamping on railway track geometry degradation. *Proceedings of the Institution of Mechanical Engineers, Part F: Journal of Rail and Rapid Transit*, **227**(4):376–391, 2013. doi: 10.1177/0954409713480439.

J. S. Bendat. *Nonlinear System Techniques and Applications*. John Wiley & Sons, Inc., 2nd edition, 1998. ISBN: 9780471165767.

A. R. B. Berawi, R. Delgado, R. Calcada, and C. Vale. Evaluating track geometrical quality through different methodologies. *International Journal of Technology*, **1**:38–47, 2010.

S. Bhaduri. *Algorithm to enable intelligent rail break detection*. PhD thesis, Virginia Polytechnic Institute and State University, 2013. https://theses.lib.vt.edu/theses/available/etd-12242013-094021/unrestricted/Bhaduri_S_T_2013_1.pdf.

J. Bremsteller. High Speed Grinding, 2014. ISSN: 10417958. http://www.szdc.cz/soubory/konference-a-seminare/zdc-2014/b03-bremsteller-vossloh.pdf.

S. Bruni, I. Anastasopoulos, S. Alfi, A. Van Leuven, and G. Gazetas. Effects of train impacts on urban turnouts: modelling and validation through measurements. *Journal of Sound and Vibration*, **324**(3):666–689, 2009. doi: 10.1016/j.jsv.2009.02.016.

D. F. Cannon, K.-O. Edel, S. L. Grassie, and K. Sawley. Rail defects: an overview. *Fatigue & Fracture of Engineering Materials & Structures*, **26**(10):865–886, 2003. doi: 10.1046/j.1460-2695.2003.00693.x.

G. Carr. Embracing Technology for Railroad Track Inspection, 2014. http://railtec.illinois.edu/cee/William W. Hay/Spring2014presentations/HaySeminar2014_GCarr_Final.pdf.

H. Chang, R. Liu, and Q. Li. A multi-stage linear prediction model for the irregularity of the longitudinal level over unit railway sections. In *Computers in Railways XII*, 2010. ISSN 17433509.

G. Charles, R. Goodall, and R. Dixon. Vehicle system dynamics model-based condition monitoring at the wheel-rail interface. *Vehicle System Dynamics*, **46**:415–430, 2008. doi: 10.1080/00423110801979259.

J. Choi. *Qualitative analysis for dynamic behavior of railway ballasted track*. PhD thesis, TU Berlin, 2014.

S. Chrismer and E. T. Selig. Computer model for ballast maintenance planning. *Proceedings of 5th International Heavy Haul Railway*, 1993. https://scholar.google.com/scholar_lookup?title=Computer%20model%20for%20ballast

%20maintenance%20planning&author=S%20Chrismer&author=ET%20Selig& publication_year=1993#%230.

P. A. Cuervo, J. F. Santa, and A. Toro. Correlations between wear mechanisms and rail grinding operations in a commercial railroad. *Tribology International*, **82**:265–273, 2015. doi: 10.1016/j.triboint.2014.06.025.

T. Dahlberg. Some railroad settlement models-a critical review. *Proceedings of the Institution of Mechanical Engineers, Part F: Journal of Rail and Rapid Transit*, **215**(4):289–300, 2001. doi: 10.1243/0954409011531585.

D. Einbinder. *The development of rail defects due to the presence of geometry defects in class 1 railroads*. Master thesis, University of Delaware, 2015. http:// dspace.udel.edu/bitstream/handle/19716/17394/2015_EinbinderDaniel_MCE .pdf?sequence=1&isAllowed=y.

H. Elaina. Track engineering standard rail grinding. Technical report, Metro. MTST 033100-08, 2015.

C. Esveld. *Modern Railway Track*. MRT-Productions, 2001. ISBN: 9080032433.

FRA. Track safety standards: improving rail integrity. Technical report, Federal Railroad Administration, 2014a. https://www.fra.dot.gov/eLib/Details/L04920.

FRA. Automated track inspection program (ATIP) geometry car operation. In *Track and Rail Infrastructure Integrity Compliance Manual*, Chapter 3, pages 1–22. Federal Railroad Administration, 1st edition, 2014b.

FRA. Track inspector rail defect reference manual. Technical report, Federal Railroad Administration, 2015. https://www.fra.dot.gov/Elib/Details/L03531.

S. Galvan-Nunez. *Hybrid Bayesian-Wiener Process in Track Geometry Degradation Analysis*. Ph.D. Thesis, 2017.

X. Gibert, V. M. Patel, and R. Chellappa. Deep Multi-Task Learning for Railway Track Inspection, 2015. http://arxiv.org/abs/1509.05267.

H. Guler. Prediction of railway track geometry deterioration using artificial neural networks: a case study for Turkish state railways. *Structure and Infrastructure Engineering*, **10**(5):1–13, 2013. doi: 10.1080/15732479.2012.757791.

A. Haigermoser, B. Luber, J. Rauh, and G. Gräfe. Road and track irregularities: measurement, assessment and simulation. *Vehicle System Dynamics*, **53**(7):878–957, 2015. doi: 10.1080/00423114.2015.1037312.

A. Hamid, K. Rasmussen, M. Baluja, and T.-L. Yang. Analytical descriptions of track geometry variations. Technical report, Report No. DOT/FRA/ORD-83/ 03.1, 1983.

C. M. Hartsough, J. W. Palese, G. Schmitzer, J. C. Espindola, and T. G. Viana. Optimized rail grinding through dynamic positioning and powering of grinding motors. In *2016 Joint Rail Conference*, page V001T01A005. ASME, 2016. ISBN: 978-0-7918-4967-5.

W. J. Harris, W. Ebersöhn, J. Lundgren, H. Tournay, and S. Zakharov, *Guidelines to Best Practices for Heavy Haul Railway Operations: Wheel and Rail Interface Issues*, International Heavy Haul Association, 2001.

W. W. Hay. *Railroad Engineering*. Wiley, 1982. ISBN: 9780471364009.

Q. He, H. Li, D. Bhattacharjya, D. P. Parikh, and A. Hampapur. Railway track geometry defect modeling: deterioration, derailment risk, and optimal repair. In *Transportation Research Board 92nd Annual Meeting*, pages 1–20, 2013. http://trid.trb.org/view.aspx?id=1242877.

R. Huber-Mörk, M. Nölle, A. Oberhauser, and E. Fischmeister. Statistical rail surface classification based on 2D and 21/2D image analysis. In *Advanced Concepts for Intelligent Vision Systems*, Volume **6474**, pages 50–61. Springer, Berlin Heidelberg, 2010. http://link.springer.com/10.1007/978-3-642-17688-3_6.

P. A. Y. Ibrekk. *Detecting anomalies and water distribution in railway ballast using GPR*. Master thesis, Norwegian University of Science and Technology, 2015. https://brage.bibsys.no/xmlui/handle/11250/2381904.

R. N. Iyengar and O. R. Jaiswal. Random field modeling of railway track irregularities. *Journal of Transportation Engineering*, **121**:303–308, 1995. doi: 10.1061/(ASCE)0733-947X(1997)123:3(245.2).

S. Jovanovic. Railway track quality assessment and related decision making. In *AREMA 2006 Annual Conferences*, 2006.

A. D. Kerr. *Fundamentals of Railway Track Engineering*. SimmonBoardman Books, Inc., 2003. ISBN: 0911382402, 9780911382402.

I. A. Khouy, H. Schunnesson, U. Juntti, A. Nissen, and P.-O. Larsson-Kråik. Evaluation of track geometry maintenance for heavy haul railroad in Sweden - a case study. *Proceedings of the Institution of Mechanical Engineers*, Part F: Journal of Rail and Rapid Transit, 496–503, 2013. doi: 10.1177/0954409713482239.

R.-K. Liu, P. Xu, Z.-Z. Sun, C. Zou, and Q.-X. Sun. Establishment of track quality index standard recommendations for Beijing metro. *Discrete Dynamics in Nature and Society*, **2015**:1–9, 2015. ISSN 1026-0226. doi: 10.1155/2015/473830. URL http://www.hindawi.com/journals/ddns/2015/473830/.

M. Matsui. Application of X-Ray Fourier Analysis to Rolling Contact Fatigue Layer of Rail, 2014. http://www.rtri.or.jp/eng/publish/newsletter/pdf/48/RTN-48-283.pdf.

A. Matsumoto, Y. Sato, H. Ohno, M. Shimizu, J. Kurihara, M. Tomeoka, T. Saitou, Y. Michitsuji, M. Tanimoto, and M. Mizuno. Continuous observation of wheel/rail contact forces in curved track and theoretical considerations. *Vehicle System Dynamics: International Journal of Vehicle Mechanics and Mobility*, **50**(1):349–364, 2012. doi: 10.1080/00423114.2012.669130.

C. Meier-Hirmer, A. Senée, G. Riboulet, F. Sourget, and M. Roussignol. A decision support system for track maintenance. In *Computers in Railways X*, *WIT Transactions on The Built Environment*, Volume **1 and 88**, pages 217–226. WIT Press, Southampton, UK, 2006. ISBN: 1845641779.

S. Z. Meymand, A. Keylin, and M. Ahmadian. A survey of wheel-rail contact models for rail vehicles. *Vehicle System Dynamics: International Journal of*

Vehicle Mechanics and Mobility, **54**(3):386–428, 2016. doi: 10.1080/00423114.2015.1137956.

J. Oberg. *Track Deterioration of Ballasted Tracks: Marginal Cost Models for Different Railway Vehicles*. Rail Vehicles, Aeronautical and Vehicle engineering, Royal Institute of Technology, 2006. ISBN: 917178537X, https://books.google.com/books/about/Track_deterioration_of_ballasted_tracks.html?id=9hSUNAAACAAJ.

T. Oyama and M. Miwa. Mathematical modeling analyses for obtaining an optimal railway track maintenance schedule. *Japan Journal of Industrial and Applied Mathematics*, **23**(2):207–224, 2006. doi: 10.1007/BF03167551.

J. W. Palese, T. Euston, and A. M. Zarembski. Use of profile indices for quality control of grinding. In *Annual Conference and Exposition, AREMA*. AREMA, 2004. https://scholar.google.com/scholar?hl=en&q=Use+of+profile+indices+for+quality+control+of+grinding&btnG=&as_sdt=1%2C8&as_sdtp=.

J. W. Palese and T. W. Wright. Risk based ultrasonic rail test scheduling on Burlington Northen Santa Fe. In *AREMA Proceedings of the 2000 Annual Conference*, pages 1–35, 2000. https://trid.trb.org/view.aspx?id=675467.

M. Papaelias, C. Roberts, and C. L. Davis. A review on non-destructive evaluation of rails: state-of-the-art and future development. *Proceedings of the Institution of Mechanical Engineers, Part F: Journal of Rail and Rapid Transit*, **222**(4):367–384, 2008. doi: 10.1243/09544097JRRT209.

L. M. Quiroga and E. Schnieder. Monte Carlo simulation of railway track geometry deterioration and restoration. *Proceedings of the Institution of Mechanical Engineers, Part O: Journal of Risk and Reliability*, **226**(3):274–282, 2012. doi: 10.1177/1748006X11418422.

S. Ramírez-Gallego, S. García, H. Mouriño-Talín, D. Martínez-Rego, V. Bolón-Canedo, A. Alonso-Betanzos, J. M. Benítez, and F. Herrera. Data discretization: taxonomy and big data challenge. *Wiley Interdisciplinary Reviews: Data Mining and Knowledge Discovery*, **6**(1):5–21, 2016. doi: 10.1002/widm.1173.

M. Rose. Shaping the Future of Rail Through Technology, 2015.

J. Sadeghi. Development of railway track geometry Indexes based on statistical distribution of geometry data. *Journal of Transportation Engineering*, **136**(8):693–700, 2010. doi: 10.1061/(ASCE)0733-947X(2010)136:8(693).

J. Sadeghi and H. Askarinejad. Development of improved railway track degradation models. *Structure and Infrastructure Engineering*, **6**(6):675–688, 2010. doi: 10.1080/15732470801902436.

J. C. Santamarina and D. Fratta. *Introduction to Discrete Signals and Inverse Problems in Civil Engineering*. ASCE Press, 1998. https://books.google.com/books/about/Introduction_to_Discrete_Signals_and_Inv.html?id=GIVRAAAAMAAJ&pgis=1.

Y. Sato. Japanese studies on deterioration of ballasted track. *Vehicle System Dynamics*, **24**(sup1):197–208, 1995. doi: 10.1080/00423119508969625.

S. V. Sawadisavi. *Development of Machine-vision technology for inspection of railroad track*. PhD thesis, University of Illinois at Urbana-Champaign, 2010.

E. T. Selig and J. M. Waters. *Track Geotechnology and Substructure Management*. Thomas Telford, 1994. ISBN: 0727720139.

W. Shao, A. Bouzerdoum, S. L. Phung, L. Su, B. Indraratna, and C. Rujikiatkamjorn. Automatic classification of ground-penetrating-radar signals for railway-ballast assessment. *IEEE Transactions on Geoscience and Remote Sensing*, **49**(10 Part 2):3961–3972, 2011. doi: 10.1109/TGRS.2011.2128328.

M. Shenton. Ballast deformation and track deterioration. *Track technology*, 1985.

I. Y. Shevtsov. *Wheel/rail interface optimisation*. Phd thesis, Delft University of Technology, 2008.

R. Stock. Damage in the Rail-Wheel System - An Overview, 2012. http://www .wheel-rail-seminars.com/downloads.php.

P. F. Teixeira and A. R. Andrade. Unplanned-maintenance needs related to rail track geometry. *Proceedings of the ICE - Transport*, **167**(6):400–410, 2014.

Z. Trzaska. Modeling of energy processes in wheel-rail contacts operating under influence of periodic discontinuous forces. *Journal of Transportation Technologies*, **02**(02):129–143, 2012. doi: 10.4236/jtts.2012.22014.

K. Tzanakakis. *The Railway Track and Its Long Term Behaviour, Springer Tracts on Transportation and Traffic*, Volume **2**. Springer, Berlin Heidelberg, 2013. http:// link.springer.com/10.1007/978-3-642-36051-0.

C. Vale and S. M. Lurdes. Stochastic model for the geometrical rail track degradation process in the Portuguese railway Northern Line. *Reliability Engineering and System Safety*, **116**:91–98, 2013. doi: 10.1016/j.ress .2013.02.010.

P. Veit. Track quality: luxury or necessity? Technical Report July, 2007.

P. Weston, C. Roberts, G. Yeo, and E. Stewart. Perspectives on railway track geometry condition monitoring from in-service railway vehicles. *Vehicle System Dynamics*, **3114**(7):1063–1091, 2015. doi: 10.1080/00423114.2015.1034730.

P. Xu, Q. Sun, R. Liu, and F. Wang. A short-range prediction model for track quality index. *Proceedings of the Institution of Mechanical Engineers, Part F: Journal of Rail and Rapid Transit*, **225**(3):277–285, 2011. doi: 10.1177/2041301710392477.

M. Yousefikia, S. Moridpour, S. Setunge, and E. Mazloumi. Modeling degradation of tracks for maintenance planning on a tram line. *Journal of Traffic and Logistics Engineering*, **2**(2):86–91, 2014. doi: 10.12720/jtle.2.2.86-91.

A. M. Zarembski and J. W. Palese. Characterization of broken rail risk for freight and passenger railway operations. In *2005 AREMA Annual Conference*, pages 1–25, Chicago, IL, 2005.

W.-J. Zwanenburg. Modeling degradation processes of switches and crossings for maintenance and renewal planning on the Swiss railway network. In *6th Swiss Transport Research Conference - STRC 06*, pages 1–11, Monte Verità / Ascona, 2006.

W.-J. Zwanenburg. The Swiss experience on the wear of railway switches & crossings. In *7th Swiss Transport Research Conference-STRC 07*, 2007.

2

Data Analysis – Basic Overview

2.1 Introduction

Data can be grouped into the following: (a) continuous, (b) ordinal, and (c) nominal. Continuous data are usually real numbers of quantitative measurements and have measurement errors and uncertainty associated with them. Therefore, stochastic methods are used to characterize the data. Probability distributions are commonly used to characterize this type of data. Ordinal data, or in some cases referred to as ranked variables, have discrete values which can be ordered.

Depending on the type of analysis and inference, continuous data can be arranged and sorted as ordinal data. Nominal data, which are also known as categorical data, are descriptive in nature and unordered.

Most of the data in railway track engineering are contained in the form of relational database management and are either in the form of Structured Query Language (SQL) or non-SQL management systems. There are a few examples of this, including Track Data Management System (TDMS), which is the collection of geometry data that can be used, among other things, to perform basic statistical analysis and visualization of the defects.

2.2 Exploratory Data Analysis (EDA)

Data analysts use exploratory statistics and, in some cases, basic data mining techniques to explore the data before any formal analysis and modeling. Exploratory data analysis (EDA) looks at the data in different ways to detect or explore interesting features, including anomalies in the data. In most cases, the EDA does not have a set of standard mathematical or statistical tools, but this initial analysis output is heavily dependent on the analysts, understanding of various statistical analysis methods, machine learning techniques, and other data analysis algorithms. Basic information, like the distribution of the

Big Data and Differential Privacy: Analysis Strategies for Railway Track Engineering, First Edition. Nii O. Attoh-Okine.
© 2017 John Wiley & Sons, Inc. Published 2017 by John Wiley & Sons, Inc.

data, measuring the center, and spread, is very critical. In traditional data analysis, the line between EDA and statistics is very blurred (Morgenthaler, 2009). Morgenthaler (2009) listed some typical EDA procedures. These include (a) the median, which appears to be more resistant than the average, and (b) the box plots, which are based on the five-number summary. This summary is made up of a branch of numbers by two extreme values (E), the two hinges (H), and the median (M). The four intervals indicate the 0–25, 25–50, 50–75, 75–100 percentiles. In some cases, the plots can show the presence of outliers in the data. Figure 2.1 shows a box plot of some track geometry parameters.

Figure 2.2 shows distribution of selected track geometry parameters.

Another important EDA is residual plots. For example, the analysis can compare the box plot of the residual with the original data. If the spread of the box plot is reduced compared to the original data, then one can initially conclude that most of the information in the original data has been taken into account, although this may not be the final conclusion. The running median appears to another EDA technique which is very effective (Morgenthaler, 2009). The quantile also appears to be another EDA analysis which can reinforce the normality of the data and the extremes, if any, present in the data. The quantile is used to

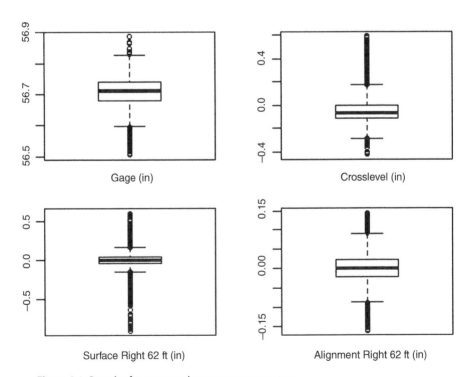

Figure 2.1 Box plot for some track geometry parameters

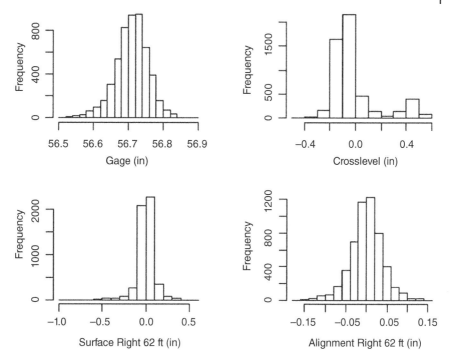

Figure 2.2 Histogram for some track geometry parameters

plot a Q–Q plot, a standard normal distribution of the form. Figure 2.3 shows Q–Q plot of selected geometry parameters.

Figure 2.4 shows different sections of track surface inspection data collected at different time periods. The structure and properties of series visually depict different kinds of behavior and maintenance decisions. Figure 2.5 shows some basic scatter plots at different time periods.

$$
y_i = \begin{cases}
x_i, & \text{if } x_i = x_{i+1} \text{ and } x_i = x_{i-1} \\
x_i, & \text{if } x_i = x_{i+1} \text{ and } x_i = x_{i+2} \\
\text{median}(x_{i-1}, x_i, x_{i+2}), & \text{if } x_i = x_{i+1} \text{ and } x_i \neq x_{i-1} \text{ and } x_i \neq x_{i+2} \\
\text{median}(x_{i-2}, x_i, x_{i+1}), & \text{if } x_i \neq x_{i+1} \text{ and } x_i = x_{i-1} \\
\text{median}(x_{i-1}, x_i, x_{i+1}), & \text{if } x_i \neq x_{i+1} \text{ and } x_i = x_{i-1}.
\end{cases}
$$

(2.1)

The running median has $y_2 = \text{median}(x_1, x_2, x_3)$.

$$
y_2 = \begin{cases}
x_2, & \text{if } x_1 = x_2 \\
\text{median}(x_1, x_2, x_4), & \text{if } x_1 \neq x_2 \text{ and } x_2 = x_3 \\
\text{median}(x_1, x_2, x_3), & \text{if } x_1 \neq x_2 \text{ and } x_2 \neq x_3.
\end{cases}
$$

(2.2)

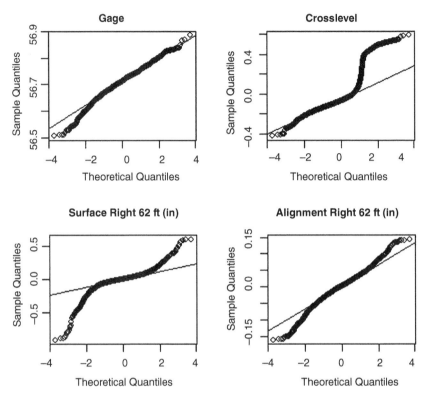

Figure 2.3 Q–Q plot for some track geometry parameters

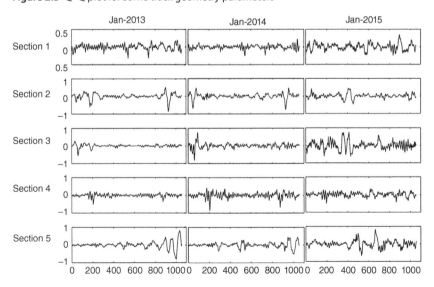

Figure 2.4 Time series data table – track surface inspection

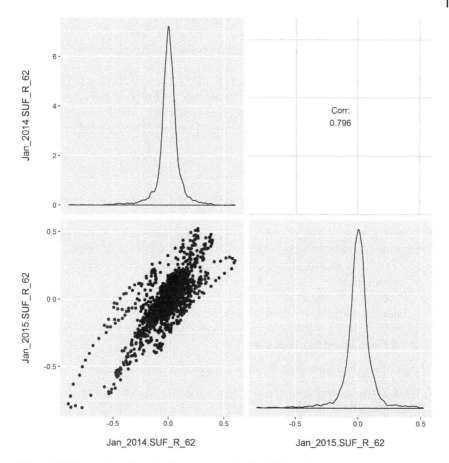

Figure 2.5 Illustration of multivariate scatter plots for different inspection times

In some cases, some observations may have different observations from the majority data. Such observations are known as outliers (Rousseeuw and Hubert, 2011). When they do affect the results, they are known to have a masking effect, but if this happens to be good data, then it is swamping.

2.3 Symbolic Data Analysis

Symbolic data analysis (SDA) provides a framework for the representation and analysis of railway track data that comprehends inherent variability. There are two different types of SDA (Brito, 2014). (a) **Temporal aggregation** is when the data being detected are collected at different times for the same entities, but time is not an issue. In these cases, the data is aggregated so that the whole set is considered. (b) **Contemporary aggregation** is the case when data are recorded at the same point in time, but for different purposes at a higher level than that for which the data was originally collected. Table 2.1 is an example of symbolic data.

There are various techniques used in SDA. These include the following: (a) Clustering – Various types of clustering techniques have been used. (b) Classification – Especially for interval-valued data. (c) Principal component analysis (PCA). (d) Copula methods – These will be discussed in the later chapters. (e) Galois lattices approach.

2.3.1 Building Symbolic Data

- *First step*. A standard data table is used where subdivisions are described by numerical categorical random variables Y_j.
- *Second step*. A table is used/created where classes of subdivisions are described by random variables Y'_j with random variables Y_{ij} value.
- *Third step*. A symbolic data table is obtained where random variables Y_{ij} are represented by:
  - Probability distributions, histograms, bar charts, and percentiles
  - Intervals Min, Max, interquartile interval, and so on.

2.3.2 Advantages of Symbolic Data

Diday (2014) highlights the advantages of symbolic data as follows:

- Work at the needed level of generality without losing variability.
- Reduce simple or complex huge data.
- Reduce number of observations and number of variables.
- Reduce missing data.
- Extract simplified knowledge and decisions from complex data.
- Facilitate interpretation of results: decision trees, factorial analysis, and new graphic kinds.

2.4 Imputation

Missing data is common in track databases; this may be due to many factors, including (a) equipment malfunctioning and (b) operator errors. In order to address missing data issues and have statistically valid inferences Rässler et al. (2013), three criteria have to be satisfied: (a) approximate unbiased estimates of population estimates, (b) interval estimates with at least nominal coverage, and (c) tests of significance should reject at their nominal level or less frequency when the null hypothesis is true.

Table 2.1 Examples of symbolic variables.

Kind of variable	Single	Multivalued	Numerical modal	Categorical modal	Interval modal
Numerical	Standard numerical variable	Symbolic numerical multivalued variable	Symbolic (number, number) list valued variable		
Categorical	Standard categorical variable	Symbolic categorical multivalued variable	Symbolic (number, category) list valued variable	Symbolic (category, category) list valued variable	
Interval	Symbolic interval-valued variable	Symbolic interval multivalued variable	Symbolic (number, interval) list valued variable	Symbolic (category, interval) list valued variable	Symbolic (interval, interval) list valued variable

These are some of the examples of handling missing data: (a) conventional methods, which involve listwise deletion (removing cases that have missing variables, and (b) imputation methods, which involve substituting each missing value for a reasonable guess and then proceeding to analysis, assuming there were no missing values.

The two main imputation techniques are (a) marginal mean imputation, where the mean of the variable is computed using non-missing values and used to impute the missing values, which can have a biased estimate on the variances and covariances, and (b) conditional mean approach, using regression to predict the missing values. There are a few advances in methods, which include multiple imputation, maximum likelihood methods, and Bayesian simulation methods (Soley-Bori, 2013).

2.5 Bayesian Methods and Big Data Analysis

The current big data era offers multiple sources of track monitoring and testing data that contain a variety of information in extremely large volumes. For example, rail defect images and the combination of geometry data and subsurface data will allow better assessment of the rail track safety and reliability. Most of the data in railway track engineering will not have the heterogeneity that makes Bayesian methods very applicable. Much of the limited statistical models used in railway track data analysis have been focused on fitting models with limited data; now the problem is rather the opposite.

Although the Markov Chain Monte Carlo (MCMC) has been successfully used as an effective Bayesian inference method, the growth and different variety of data in railway track applications may not have direct applications as used in previous MCMC. The big data might have a case with large $p \geq N$. In this case, the use of traditional statistical methods may not be applicable. For large N, a subset of the data can be used in the MCMC framework. Data with a large number of observations is often referred to as tall data (Baker et al., 2015). Two categories of data analysis are discussed:

- *Batch methods.* This method splits the data into different subsets. The idea of batch methods for big data is to run MCMC separately to sample from each sub-posterior. The samples are then combined, and the final sample will follow the full posterior as close as possible. The sub-posteriors can be combined using parametric methods (normal distribution), nonparametric, and semi-parametric methods.
- *Stochastic gradient methods.* This uses optimization-based tools. The aim is to optimize the likelihood function. Only each subset of the data is used for the optimization. The model used by this approach can suffer from

overfitting since it does not sample from the posterior like the traditional MCMC. It has a tendency to become stuck at the local maxima. Some of the methods used include stochastic gradient Langevin dynamics (SGLD) and stochastic gradient Hamiltonian Monte Carlo (SGHMC).

The specification of the prior has been a major criticism of Bayesian analysis, but in large-scale data, the specification of the prior can be a major advantage. Therefore, convergence can be faster with an informative prior. Therefore, as the number data points increase, the prior becomes more peaked. Oravecz et al. (2015) proposed sequential Bayesian updating for big data, which tends to address the volume and velocity.

There are two general approaches of Bayesian inference: (a) MCMC sampling and variational optimization. The MCMC is asymptotically unbiased and suffers from significant variance. (b) The variational optimization is deterministic and does not have variance but can exhibit biases Yeh et al. (2012).

2.6   Remarks

Basic data analysis is still very important for initial track analysis. It gives insight into the data properties and, hence, can provide a better direction for any statistical assumption needed for future analysis. Furthermore, there are some inspection data in track monitoring which are still not the realm of big data.

References

J. Baker, P. Fearnhead, and J. Baker. Computational statistics for big data. Technical report, Lancaster University, 2015. http://www.lancaster.ac.uk/pg/bakerj1/pdfs/big_data/big_data.pdf.

P. Brito. Symbolic data analysis: another look at the interaction of Data Mining and Statistics. *Wiley Interdisciplinary Reviews: Data Mining and Knowledge Discovery*, 4(4):281–295, 2014. doi: 10.1002/widm.1133.

E. Diday. Introduction to Symbolic Data Analysis, 2014. http://www.docfoc.com/introduction-to-symbolic-data-analysis-e-diday-ceremade-parisdauphine.

S. Morgenthaler. Exploratory data analysis. *Wiley Interdisciplinary Reviews: Computational Statistics*, 1(1):33–44, 2009. doi: 10.1002/wics.2.

Z. Oravecz, M. Huentelman, and J. Vandekerckhove. Sequential Bayesian Updating for Big Data, 2015. http://www.cidlab.com/prints/oravecz2016sequential.pdf.

S. Rässler, D. B. Rubin, and E. R. Zell. Imputation. *Wiley Interdisciplinary Reviews: Computational Statistics*, 5(1):20–29, 2013. doi: 10.1002/wics.1240.

P. J. Rousseeuw and M. Hubert. Robust statistics for outlier detection. *Wiley Interdisciplinary Reviews: Data Mining and Knowledge Discovery*, **1**(1):73–79, 2011. doi: 10.1002/widm.2.

M. Soley-Bori. Dealing with missing data: key assumptions and methods for applied analysis. Technical Report 4, Boston University. School of Public Health, 2013.

Y.-T. Yeh, L. Yang, M. Watson, N. D. Goodman, and P. Hanrahan. Synthesizing open worlds with constraints using locally annealed reversible jump MCMC. *ACM Transactions on Graphics*, **31**(4):1–11, 2012. doi: 10.1145/2185520 .2185552.

3

Machine Learning: A Basic Overview

3.1 Introduction

Machine learning (ML) generally refers to the development of methods that optimize their performance iteratively by "learning from the data." For example, let us assume the detection of geometry defects of railway tracks as a general problem of ML. The ML formulation can be as follows:

Find $f : \mathbf{x} \rightarrow y = f(\mathbf{x})$, where $\mathbf{x}$ represents the geometry defects and y is a scalar function that indicates the presence ($y = 1$) or absence ($y = 0$) of geometry defects.

Depending on the type of training data, assumptions of various approaches can be used:

- Supervised learning
- Unsupervised learning
- Semi-supervised learning

Furthermore, $f(\mathbf{x})$ can either be:

- Linear algorithms – a linear relationship between $\mathbf{x}$ and y
- Nonlinear algorithms – the relationship between $\mathbf{x}$ and the label y is nonlinear and f can be any function

  The nonlinear approach can be further classified as parametric or nonparametric. In the parametric cases, the parameters to be estimated are assumed to follow a specific distribution. The nonparametric approaches do not make any prior assumptions on the distribution of input data. Kernel methods are typically used for this approach. The kernel functions are of the form

$$f(\mathbf{x}) = \sum_{i=1}^{\ell} \alpha_i K\left(\mathbf{x}_i, x\right) + b, \tag{3.1}$$

where K is the kernel function and $((\alpha_{i=1}^{\ell}), b)$ are parameters of the function.

Big Data and Differential Privacy: Analysis Strategies for Railway Track Engineering, First Edition. Nii O. Attoh-Okine.
© 2017 John Wiley & Sons, Inc. Published 2017 by John Wiley & Sons, Inc.

ML is broadly understood as a group of methods that analyze data and make useful discoveries and inferences from the data. It relies heavily on techniques and theory from statistics, optimization, algorithms, and biologically inspired systems. ML can be classified into four groups:

- Supervised learning
- Unsupervised learning
- Semi-supervised learning
- Reinforcement learning

3.2 Supervised Learning

This is the learning technique that handles training data labeled with a desired output to guide learning process. Mathematically, it can be addressed as follows:

Let training set $X =$ equation composed of pairs of input and output variables. It is assumed that there is a dependency between them, namely, that the output is some unknown function of the input and some other unobservable variables z^t.

$$\text{Using } (\mathbf{x}_i, y_i)_{i=1}^{\ell}, \quad \text{find } f : \mathbf{x} \mapsto y = f(\mathbf{x}). \tag{3.2}$$

The objective is to fit a model to learn the mapping from the observable input to the output

$$y^t = g(\mathbf{x}^t \mid \theta), \tag{3.3}$$

where $g(\cdot)$ is the model and θ denotes its parameters. Learning corresponds to finding the parameters that minimize an error function by measuring the deviation of the prediction y^t from the desired output r^t, as given by a loss function:

$$\arg \min_{\theta} \sum_t L(r^t, y^t) = \arg \min_{\theta} \sum_t L\left(r^t, g\left(\mathbf{x}^t \mid \theta\right)\right). \tag{3.4}$$

The loss can be a fraction of the incorrect predictions.

Different supervised learning algorithms differ in the models or the loss functions they assume or in the procedure they use in optimization. If the output r^t is a discrete label, this is a classification problem, $g(\cdot)$ is a regressor, and loss is the squared error. The minimization to find the best θ can be easy or hard depending on the model and the loss (Alpaydın, 2011). The main goal of supervised learning is to learn a model that predicts an output given a new input. In non-deterministic cases, the probability distribution between the input "features" x and the corresponding output label y will be $p(y \mid x)$.

3.3   Unsupervised Learning

This is the type of learning where examples are not labeled. This includes clustering and dimensionality reduction. For example, clustering outputs can be used to extract key concepts from the data set. It can be very subjective. The training set $X = $ equation has only the set of variables drawn from some unknown probability density,

$$\mathbf{x}^t \sim p(\mathbf{x}), \tag{3.5}$$

and the aim is to estimate it using a model with parameters θ

$$\mathbf{x}^t \sim q(\mathbf{x} \mid \theta). \tag{3.6}$$

Again, a value of θ makes q as close as possible to the unknown p. And again, because $p(\cdot)$ is unknown and only has a sample drawn from it, similarity is measured on this representative sample. In this case, a θ that maximizes the probability of drawing the sample is sought; this is called maximum likelihood estimation:

$$\arg \max_{\theta} \prod_{t} q(\mathbf{x}^t \mid \theta). \tag{3.7}$$

A variety of models can be assumed for $q(\cdot)$: When x^t is believed to come from one of a small number of groups, it is a clustering application (Alpaydın, 2011).

3.4   Semi-Supervised Learning

The learning methods come from a collection of cases where only parts of them have assigned labels.

3.5   Reinforcement Learning

This is an approach that controls dynamic programming and supervised learning. There is no training set; rather the learning system generates data by itself. The objective is to learn a policy, that is, a model that the system uses to generate a sequence of outputs so as to maximize the cumulative reward.

$\hat{Q}(s^t, a^t)$ denotes the cumulative reward expected starting from state s^t and taking action at a^t time t. The idea is to estimate using a model and the parameters

$$\hat{Q}(s^t, a^t) = g(s^t, a^t \mid \theta). \tag{3.8}$$

These values for any intermediate state are unknown or are received after carrying out a whole sequence of actions called an episode. Bellman's equation is written as follows:

$$Q^*(s^t, a^t) = E[r^{t+1}] + \gamma \sum_{s^{t+1}} P\left(s^{t+1} \mid s^t, a^t\right) \max_{a_{t+1}} Q^*\left(s^{t+1}, a^{t+1}\right). \qquad (3.9)$$

The first term to the right is the expected immediate reward, and the second term is the discounted expected future reward where $\gamma < 1$ is the discount factor. For all possible next states $s^{(t+1)}$, probability $P(s^{t+1} \mid s^t, a^t)$ is used and assuming that the best action is taken, the expected reward thereafter is $\max_{a_{t+1}} Q*(s^{t+1}, a^{t+1})$ (Alpaydın, 2011). Selected examples of algorithms for ML will be discussed later in the chapter. The assessment of ML algorithms is very critical; in some cases, one has to set a particular algorithm based on their performance. The performance is done in three stages:

1) Training set – used to optimize the parameters of a given model.
2) Validation set – used to optimize the model. This is known as model selection.
3) Test set – used to assess the accuracy of the final model.

Different methods are used for selecting these sets: resampling, cross-validation, and bootstrap (Alpaydın, 2011). In general, ML has different meanings depending on the application. Supervised learning for classification is referred to as pattern recognition; data may when ML methods are used to

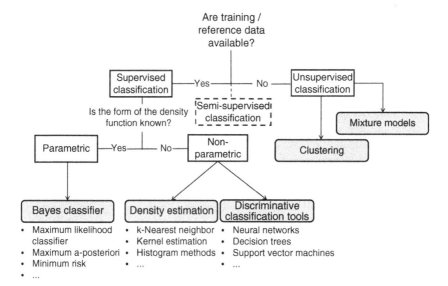

Figure 3.1 Overview of different classifier categories (Camps-Valls and Bruzzone (2009). Reproduced with the permission of John Wiley and Sons)

make inferences in large data. If the objective is the prediction of new cases or knowledge discovery, then the approach is referred to as knowledge discovery in databases. Figure 3.1 is a general overview of different classifier categories.

3.6 Data Integration

Data integration methods can be divided into (a) *homogeneous*, where the same type of data, but across multiple perspectives (e.g., experimental studies), and (b) *heterogeneous*, where multiple data types in different formats are integrated. The latter is computationally more challenging, since it requires a framework that can deal with heterogeneous data without transforming it. There is a tendency to lose information during the transformation.

3.7 Data Science Ontology

Figure 3.2 shows the ontology of data science with an example.

All these groups can be further divided into subgroups in a hierarchical manner. Some examples are deep learning, kernel methods, clustering, decision trees, and probabilistic classifiers.

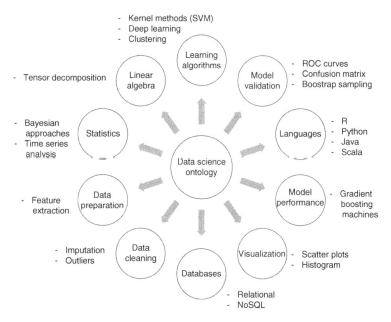

Figure 3.2 Illustration of data science ontology

3.7.1 Kernels

3.7.1.1 General

Processing railway track data is becoming more complex. For example, most of the image classification in railway track monitoring involves feature extraction. The characteristics of the images acquired during track monitoring are analyzed, for example, by using traditional neural networks. The traditional neural networks are usually affected in terms of performance by the high input dimensions, which lead to a relatively poor number of samples.

Kernel methods are an example of the ML techniques that are very effective for feature extraction. Kernel methods use kernel techniques to implicitly map input patterns to a feature space.

Kernel methods work by embedding data items (track geometry, rail defects, subsurface data) into a vector space F, called a feature space, and searching for linear relations in such a space. The embedding is defined through the use of the inner product for the feature space through a semidefined kernel function:

$$k(x_1, x_2) = \langle \Phi(x_1), \Phi(x_2) \rangle, \tag{3.10}$$

where $\Phi(x_1)$ and $\Phi(x_2)$ are the embedding of data items x_1 and x_2.

Evaluating the kernels on all pairs of data items yields a symmetric position-definite matrix k known as a kernel matrix. This forms a generalized similarity measure among the data points (Scholkopf et al., 2004).

A kernel is a function k that for all $x_1, x_2 \in x_2$ satisfies

$$k(x_1, x_2) = \langle \phi(x_1), \phi(x_2) \rangle, \tag{3.11}$$

The only information used from the training examples is their so-called kernel matrix, defined as the square matrix $\mathbf{K} \in \mathbb{R}^{\ell \times \ell}$ such that $\mathbf{K}_{ij} = k(x_i, x_j)$ for a set of vectors $\{x_1, \dots, x_\ell\} \subseteq X$ and some kernel function k.

3.7.1.2 Learning Process

Lodhi (2012) presented the learning process of the kernel methods:

- *Stage 1.* Data is mapped into a higher dimensional space through some nonlinear mapping Φ. The mapped space is referred to as the feature space F, and the mapping is given by

$$\Phi : \Omega \to F. \tag{3.12}$$

- *Stage 2.* The mapping Φ may not be known. Ω is the input space. The kernels measure the similarity by calculating the inner product between the images of the examples in the (high-dimensional) feature space. Mathematically, the feature mapping can be expressed as follows:

$$\Phi : \Omega \to H$$
$$x_1, \dots, x_n \to \Phi(x_1), \dots, \Phi(x_n). \tag{3.13}$$

Depending on the special choice of feature space and kernel trick,

$$\langle \Phi(x_i), \Phi(x_j) \rangle = k(x_i, x_j). \tag{3.14}$$

The matrix obtained by computing the inner product between n-examples is known as a Gram matrix or kernel matrix and contains all the information required by kernel-based methods.

3.7.2 Basic Operations with Kernels

- *Translational in Feature Space*

$$\widetilde{\Phi}(x) = \Phi(x) + \Gamma \tag{3.15}$$

Translational dot product

$$\left\langle \widetilde{\Phi}(x_i), \widetilde{\Phi}(x_j) \right\rangle \tag{3.16}$$

- *Normalize*

$$k(x_i, x_j) = \left\langle \frac{\Phi(x_i)}{\|\Phi(x_i)\|}, \frac{\Phi(x_j)}{\|\Phi(x_j)\|} \right\rangle$$

$$k(x_i, x_j) = \frac{k(x_i, x_j)}{\sqrt{k(x_i, x_j)k(x_i, x_j)}} \tag{3.17}$$

- *Computing Distance*

$$d(x_i, x_j) = \left\| \Phi(x_i), \Phi(x_j) \right\|$$

$$d(x_i, x_j) = \sqrt{k(x_i, x_i) + 2k(x_j, x_j) - 2k(x_i, x_j)} \tag{3.18}$$

3.7.3 Different Kernel Types

- Linear Kernel

$$f(\mathbf{x}) = \sum_n \alpha_i \left\langle \mathbf{x}_i, \mathbf{x}_j \right\rangle = \langle \mathbf{x}, \Sigma \alpha_i x_i \rangle = \langle \mathbf{x}, \mathbf{w} \rangle \tag{3.19}$$

$$\mathbf{w} = \sum_{i=1}^{n} \alpha_i \mathbf{x}_i \tag{3.20}$$

- Polynomial Kernel

$$k(\mathbf{x}, \mathbf{x}') = \left\langle \mathbf{x}, \mathbf{x}' \right\rangle^p, \ p \in \mathbb{N} \tag{3.21}$$

- Gaussian Kernel

$$k(\mathbf{x}, \mathbf{x}') = \exp\left(\frac{-\|\mathbf{x} - \mathbf{x}'\|^2}{2\sigma^2} \right) \tag{3.22}$$

3.7.4 Intuitive Example

For example, in railway track monitoring, data classification input patterns may be a real vector space $x \in \mathbb{R}^N$, and classification algorithms use a hyperplane in

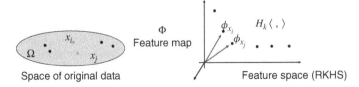

Figure 3.3 Transformation of original data to feature space

the vector space to classify the patterns. If the essential features of patterns may be all the d-order products – known as product features of dimensions of the input space – $x_{j_1}, x_{j_2}, \ldots, x_{j_d}$, $j_1, j_2, \ldots, j_s \in \{1, \ldots, N\}$. Figure 3.3 presents the transformation process.

If, for example, $N = 2$ and $d = 2$, the dimensionality N_F of the feature space F is 3:

$$(x_1, x_2) \rightarrow \left(x_1^2, x_2^2, x_1 x_2\right), \tag{3.23}$$

$$N_F = \frac{(N + d - 1)!}{d!(N - 1)!}. \tag{3.24}$$

If an image with $x = 16 \times 16$ pixels, then $N = 256$, $d = 2$, and $N_F = 10^{10}$. Such a value is impractical.

3.7.5 Kernel Methods

3.7.5.1 Support Vector Machines

A support vector machine (SVM) is a supervised learning method that is used to analyze data and recognize patterns used for classification analysis. The standard SVM is defined as a maximum margin classifier whose decision function is a hyperplane that maximally separates samples from different classes. Different linear classifiers use different criteria to find the hyperplanes. The algorithm searching for the maximum-margin hyperplane is called an SVM (Fu, 2014). The SVM only involves dot product operations on input vectors; this makes it appropriate to be kernelized. Notationally, given a labeled trained data set $\{x_i, y_i\}_{i=1}^n$.

- *Maximum Margin Classifier.* A standard SVM for classification can be mathematically expressed as follows. Using Figure 3.4, let H be a hyperplane that separates two classes of samples H_1 and H_2. The two hyperplanes are parallel to H, passing samples closest to H.
  Let the equation for the hyperplane H: be

$$H : x \cdot w + b = 0, \tag{3.25}$$

  where $x \in \mathbb{R}^n$, w is the weight coefficient vector and b is the displacement. Using two class training samples set,

$$(x, y), \quad i = 1, 2, \ldots, n, \ x \in \mathbb{R}^n, \quad y \in \{+1, -1\}, \tag{3.26}$$

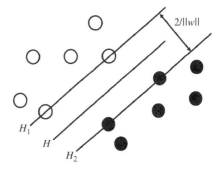

Figure 3.4 Two hyperplanes (Fu et al. (2014). Reproduced with the permission of Springer)

satisfies

$$y_i(x \cdot w + b) - 1 \geq 0 \quad i = 1, 2, \ldots, n. \tag{3.27}$$

The maximum classification margin is

$$\frac{2}{\|w\|}. \tag{3.28}$$

Maximizing the margin is equivalent to minimizing $\|w\|$.

To determine the maximum margin hyperplane, one has to solve the following optimization problem:

$$\min_{w,b} \|w\|^2 \quad \text{with} \quad y_i(x \cdot w + b) - 1 \geq 0, \quad i = 1, 2, \ldots, n. \tag{3.29}$$

The Lagrangian dual problem of this problem is

$$\max_{\alpha} \sum_{i=1}^{n} \alpha_i - \frac{1}{2} \sum_{i=1}^{n} \sum_{j=1}^{n} y_i y_j \alpha_i \alpha_j (x_i \cdot x_j) \quad \text{with} \quad \sum_{i=1}^{n} y_i \alpha_i = 0,$$

$$\forall i : \alpha_i \geq 0, \quad i = 1, 2, \ldots, n. \tag{3.30}$$

Solving this optimization problem leads to a classification function:

$$f(x) = \text{sgn}\left[\sum_{i=1}^{n} \alpha_i^* y_i (x \cdot x_i) + b^*\right]. \tag{3.31}$$

- *Cases Where the Samples Are Nonlinear.* In some cases, the data points (samples) may not be perfect and may contain noise. The SVM with the introduction of soft margins can be used to handle these issues. By inducing slack variables, the following optimization problem needs to be solved:

$$\min_{w,b} \frac{1}{2}\|w\|^2 + C \sum_{i=1}^{n} \xi_i \quad \text{with} \quad y_i(x \cdot w + b) - 1 + \xi_i \geq 0, \quad i = 1, 2, \ldots, n,$$

$$\xi_i \geq 0, \quad i = 1, 2, \ldots, n, \tag{3.32}$$

where ξ_i are the slack variables, C is the parameter controlling the trade-off between the classification margin and training errors, and the coefficient $1/2$ is introduced for representation convenience. The Lagrangian dual problem to the above problem can be stated as follows:

$$\max_{\alpha} \sum_{i=1}^{n} \alpha_i - \frac{1}{2} \sum_{i=1}^{n} \sum_{j=1}^{n} y_i y_j \alpha_i \alpha_j (x_i \cdot x_j) \quad \text{with} \quad \sum_{i=1}^{n} y_1 \alpha_i = 0, \tag{3.33}$$

$$\forall i : 0 \leq \alpha_i \leq C, \quad i = 1, 2, \dots, n.$$

The equivalent dot product is

$$\max_{\alpha} \sum_{i=1}^{n} \alpha_i - \frac{1}{2} \sum_{i=1}^{n} \sum_{j=1}^{n} y_i y_j \alpha_i \alpha_j k(x_i \cdot x_j) \quad \text{with} \quad \sum_{i=1}^{n} y_1 \alpha_i = 0, \tag{3.34}$$

$$\forall i : 0 \leq \alpha_i \leq C, \quad i = 1, 2, \dots, n.$$

The above optimization problem leads to the nonlinear soft margin classifier:

$$f(x) = \text{sgn} \left[\sum_{i=1}^{n} \alpha_i^* y_i k(x \cdot x_i) + b^* \right], \tag{3.35}$$

in which α_i^* is the solution to the problem in 3.34 and b^* can be calculated with

$$b^* = \frac{1}{y_i} - \sum_{j=1}^{n} \alpha_i^* y_j k(x_i, x_j), \tag{3.36}$$

where i is an arbitrary index satisfying $0 < \alpha_i < C$.
In more general terms,

$$b = \frac{1}{k} \sum_{i=1}^{k} \left(y_i - \langle \phi(x_i), w \rangle \right), \tag{3.37}$$

where k is the number of unbounded Lagrange multipliers and $w = \sum y_i \alpha_i \phi(x_i)$, $\alpha_i \neq 0$.

- *Training and Testing.* Training the SVM consists of an iterative process in which the SVM is given the desired inputs along with the outputs. Based on the iterative process, w and b are determined with the optimal margin. The obtained w and b are used to test unseen samples, using the trained SVM. Figures 3.5 and 3.6 show two- and three-way partitions.

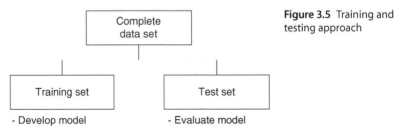

Figure 3.5 Training and testing approach

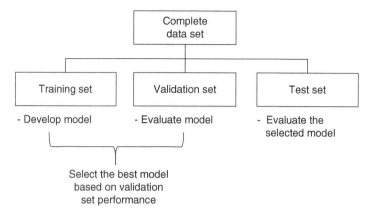

Figure 3.6 Training, testing, and validation

The general performance of SVM relies on:

a) Parameter C. This determines the trade-off between the model complexity and the degree of deviation tolerated in the optimization.
b) Kernel type.
c) Kernel parameters.

To evaluate the accuracy of the SVM, there are different variations of the SVM; these include the recovery operating characteristics (ROC), to be discussed later.

3.8 Imbalanced Classification

Imbalanced classification is a supervised learning problem where one class outnumbers the other class by an extremely large population. For example, automated inspection and monitoring rail track geometry defects or rail defects may find the number of a particular defect significantly lower than non-defective variables.

Traditional ML techniques in some cases fail to address this problem very effectively. The skewed distribution may have major influence on the classifier, biasing it toward the majority class. Methods like (a) undersampling, (b) oversampling, and (c) synthetic data generation can be used.

a) *Undersampling.* Can be random or informative. The random approach chooses observations from the majority class, until the data set is balanced, while the informative approach follows a pre-specified selection criterion.
b) *Oversampling.* Works with the minority class and replicates observations from the minority class to balance the data.

3.9 Model Validation

3.9.1 Receiver Operating Characteristic (ROC) Curves

For example, a binary classification model may classify data as true or false. This may give rise to possible classifications as follows: (a) A true positive (b) A true negative (c) A false positive (d) A false negative This can be shown as a confusion matrix (Table 3.1).

1) True positive rate $TP_{rate} = \frac{TP}{TP+FN}$ is the percentage of positive instances correctly classified.
2) True negative rate $TN_{rate} = \frac{TN}{FP+TN}$ is the percentage of negative instances correctly classified.
3) False positive rate $FP_{rate} = \frac{FP}{FP+TN}$ is the percentage of negative instances misclassified.
4) False negative rate $FN_{rate} = \frac{FN}{TP+FN}$ is the percentage of positive instances misclassified.

Alternatively, the following characteristics can be further developed:

$$Accuracy = \frac{TP + TN}{TP + TN + FP + FN} \tag{3.38}$$

$$Precision = \frac{TP}{TP + FP} \tag{3.39}$$

$$Recall = \frac{TP}{TP + FN} \tag{3.40}$$

$$Specificity = \frac{TN}{TN + FP} \quad \text{(How sensitive the classifier is to the negative cases)} \tag{3.41}$$

$$Error = \frac{FP + FN}{TP + TN + FP + FN} \tag{3.42}$$

For example, the accuracy is the percentage coverage of correct predictions to judge the predictive capabilities of the model.

Matthew correlation coefficient (MCC) varies between -1 and 1.

$$MCC = \frac{TP * TN - FN * FP}{\sqrt{(TP + FN)(TP + FP)(TN + FN)(TN + FP)}} \tag{3.43}$$

Table 3.1 Confusion matrix.

		Observed (true class)	
		True (p)	False (n)
Predicted (hypothesis class)	True (Y)	True Positive (TP)	False Positive (FP)
	False (N)	False Negative (FN)	True Negative (TN)

Figure 3.7 Receiver operating
characteristic (ROC) curve

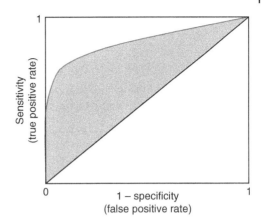

(Li et al., 2015)

$$t_p \text{ rate} = \frac{\text{Positively correctly classified}}{\text{Total positive}} \tag{3.44}$$

$$f_p \text{ rate} = \frac{\text{Negatively incorrectly classified}}{\text{Total negative}} \tag{3.45}$$

$$\{Y, N\} - \text{class prediction} \tag{3.46}$$

$$\text{Simplicity} = 1 - f_p \text{ rate} \tag{3.47}$$

(Specificity)

3.9.1.1 ROC Curves

$(0, 1)$ is perfect. $(0, 0)$, the total area of the ROC curve (Figure 3.7), is an index
for measuring the performance of the model (algorithm). The larger the AUC,
the better the overall performance. The diagonal line $y = x$ represents the strategy of randomly guessing a class (e.g., for classification). The lower triangle is
usually empty. A discrete classifier (e.g., a decision tree) produces only a single
point in the ROC.

- ROC curves
- Cross-validation

3.10 Ensemble Methods

3.10.1 General

Ensemble learning is an ML paradigm where learners are trained to solve the
same problem. It is designed to increase the accuracy of a single classifier by
training several different classifiers and combining their decisions to output a
single class label (Galar et al., 2012).

The ensemble is constructed in two steps:

1) Basic learners are produced, which can be generated in a parallel or sequential style.
2) The base learners are then combined.

The ensemble learning methods are generic and applicable across broad classes of models and learning tasks. The models can be combined using:

- Linear combiners
- Product combiners
- Voting combiners

The linear combiner is appropriate for a model that outputs real-valued numbers, for example, regression. The product combiner is used in cases with nonuniform weight like mixture of experts.

The voting method is applicable when the output is class labels. The base learners and weak learners are used in an ensemble method. Base learners are generated by base learning algorithms, for example, decision trees, neural networks, and other ML algorithms. There are quite a few ensemble methods, but only two will be discussed in this section: (a) bagging and (b) adaptive boosting (ada boosting).

3.10.2 Bagging

The bagging process begins with bootstrap resampling of the training data set. Given a data set of N pattern (track geometry or defect or a combination), a bootstrap sample is constructed by randomly sampling N patterns with replacement. Because of the replacement, some patterns may appear more than once. In Figure 3.8, instead of using one classification algorithm, different ML algorithms, decision trees, neural networks, and multiple adaptive regression splines (MARS) with different architectures are used.

In figure 3.8, many bootstrap samples first are generated from the original training set. Initially, bootstrap is based on random sampling with replacement. If, for example, the objective is to fit a linear model to our training data D, the prediction $f(x)$ at input x is obtained. N samples are initially generated by randomly drawing with replacement, where $D^* = \{(x_1^*, y_1^*), (x_2^*, y_2^*), \dots, (x_N^*, y_N^*)\}$. The main idea of the aggregating approach actually means combining multiple models. For each bootstrap set $D^{*t}, t = 1, 2, \dots, T$, one model is fitted, given the prediction $f(x)$. The estimation bagging averaging the predictors over a collection of bootstrap observations can be expressed as follows (Cao et al., 2010):

$$\hat{f}_{bag}(x) = \frac{1}{T} \sum_{t=1}^{T} \hat{f}^{*t}. \tag{3.48}$$

Bagging can dramatically reduce the variance of unstable procedures like decision trees and neural networks, leading to an improved prediction. Averaging tends to reduce the variance and leaves bias unchanged.

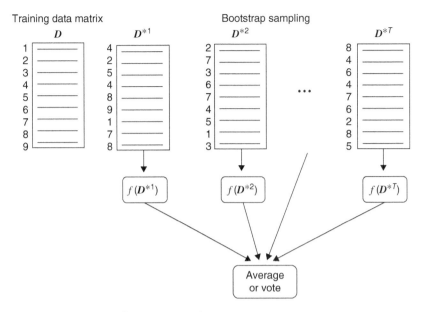

Figure 3.8 Illustration of bagging procedure

The final output, for example, in the case of regression is obtained by computing the average regression output or by majority voting for a classification problem. For example, if there are $f_1, \ldots, f_M$ classifiers. One output $\hat{y}_1$, for example, $\hat{x}$.

$$\hat{y}_1 = f_1(\hat{x})$$
$$\vdots \qquad\qquad\qquad\qquad (3.49)$$
$$\hat{y}_M = f_M(\hat{x}).$$

Algorithm 1 shows the pseudo algorithm for bagging.

Algorithm 1 Bagging

1: **Input:** Required ensemble size T
2: **Input:** Training set $S = \{(x_1, y_1), (x_2, y_2), \cdots, (x_N, y_N)\}$
3: **for** $t=1$ to T **do**
4:     Build a data set S_t by sampling N items randomly with replacement from S.
5:     Train a model h_t using S_t and add it to the ensemble.
6: **end for**
7: For a new testing point (x', y'),
8: If model outputs are continuous, combine them by averaging.
9: If model outputs are class labels, combine them by voting.

3.10.3 Boosting

Boosting is the family of algorithms. The adaptive boosting will be discussed and used for further analysis. Boosting includes an ensemble of classifiers by adaptive changing of the distribution of the training set based on previous accuracy of the previously created classification.

Algorithm 2 shows the AdaBoost algorithm.

Algorithm 2 AdaBoost Algorithm (Kotsiantis, 2013)

1: **Input:** $(x_1, y_1), \cdots, (x_m, y_m); x_i \in \chi, y_i \in \{-1, +1\}$
2: Initialize weights $D_1(i) = 1/m$
3: **for** $t=1$ to T **do**
4:      Find $h_t = \arg\min_{h_j} \epsilon_j = \sum_{i=1}^m D_t(i)(y_i \neq h_j(x_i))$
5:      **if** $\epsilon_t \geq 1/2$ **then**
6:          Stop
7:      **end if**
8:      Set $\alpha_t = \frac{1}{2} \log \left(\frac{1-\epsilon_t}{\epsilon_t} \right)$
9:      Update $D_{t+1}(i) = \dfrac{D_t(i) exp(-\alpha_t y_i h_t(x_i))}{Z_t}$,
         so that: $exp(-\alpha_t y_i h_t(x_i)) \begin{cases} < 1, y_i = h_t(x_i) \\ > 1, y_i \neq h_t(x_i) \end{cases}$
10:     Output the final classifier: $H(x) = sign \left(\sum_{t=1}^T \alpha_t h_t(x) \right)$
11: **end for**

The AdaBoost is constructed by iteratively adding models. Each time a model is learned, it is checked ($\epsilon_t < 0.5$) and has a performance better than randomly guessing the data. After each round, the distribution is updated to incorporate incorrectly classified samples. The D_{t+1} distribution may have been incorrectly classified in the previous models. The update causes the model to sequentially minimize the exponential bound or error rate.

3.11 Big P and Small N ($P \gg N$)

Due to the advancement of data collection technologies and the large number of geometry and rail defects, parameters can be measured in relatively few locations. The statistical analysis of this type of data can be challenging. For example, the use of statistical hypothesis testing will be invalid in such cases. Furthermore, signals can have substantiated noise in them. Also with this type of problem, some of the parameters cannot be estimated without regularization. Some of the methods that can be used to address and analyze these types of data include

Attributes (A_j)

	A_1	A_2	A_3	A_4	...	A_5	...	A_P
O_1	d_{11}	d_{12}	d_{13}	d_{14}	...	d_{1j}	...	d_{1P}
O_2	d_{21}	d_{22}	d_{23}	d_{24}	...	d_{2j}	...	d_{2P}
⋮	⋮	⋮	⋮	⋮	...	⋮	...	⋮
O_i	d_{i1}	d_{i2}	d_{i3}	d_{i4}	...	d_{ij}	...	d_{iP}
⋮	⋮	⋮	⋮	⋮	...	⋮	...	⋮
O_N	d_{N1}	d_{N2}	d_{N3}	d_{N4}	...	d_{Nj}	...	d_{NP}

Objects (O_i)

d_{ij}: data point associated with the object i and attribute j

Figure 3.9 Big P and small N

- The use of support vector classifiers
- Regularized discriminant analysis

Figure 3.9 depicts the concept of big P and small N $(P \gg N)$

3.11.1 Bias and Variances

Most of the learning algorithms use mathematical or statistical models. The "errors" from the models have two main components (Gutierrez, 2014):

a) Reducible error
b) Irreducible error (inherent uncertainty)

The reducible error can be decomposed into "error due to squared bias" and "error due to variance." The error due to squared bias is the amount expected when the model prediction differs from the true target value for the training data. This is due to the model selection. The error due to variance is the difference between the predictions over one training data set and the experimental/field value. The variance therefore measures how inconsistent predictions are from one another over different training sets. Models that exhibit small variance and high bias underfit the truth target models that exhibit high variance, and models with low bias overfit the truth target.

To address the bias–variance trade-off, there are different approaches to analysis that will need to use both training and test error. Figure 3.10 shows bias/variance decomposition method. Figure 3.11 shows the bias/variance methods for the appropriate learning algorithm.

3.11.2 Multivariate Adaptive Regression Splines (MARS)

Friedman (1991) introduced the MARS approach as a method for fitting the relationship between a set of predictors and dependent variables. MARS is

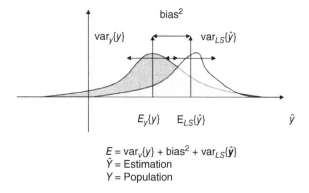

Figure 3.10 Bias/variance decomposition

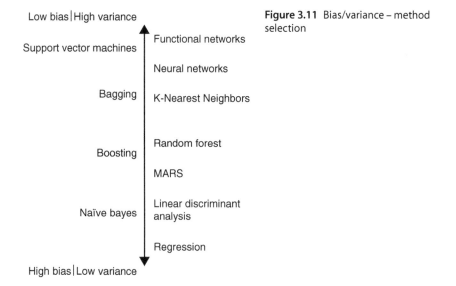

Figure 3.11 Bias/variance – method selection

based on a divide-and-conquer strategy, partitioning the training data sets into separate regions, each of which receives its own regression line (Steinberg, 2001). MARS is a data-driven procedure.

The basic problem facing railway track engineers in performance and deterioration modeling is how best to determine the fundamental relationship between the dependent variable and a vector of predictors (surface distress, age, traffic, environmental factors). The question is how best to specify f in the following equation:

$$\text{MGT} = f(G_1, G_2, \dots, G_n), \tag{3.50}$$

where $\boldsymbol{G} = \{G_1, G_2, \dots, G_n\}$ is a vector of parameters.

The MARS algorithm searches over all possible knot locations, and across interactions among all variables. It does so through the use of combinations of variables called basis functions (Sephton, 2001). The approach is analogous to the use of splines. An extremum point t, called the knot location, is identified, and two linear segments describe the region to the right (Equation 3.51) and to the left (Equation 3.52) of the given knot location, t:

$$b^+(x \mid t) = +(x - t)^q_+. \tag{3.51}$$
$$b^-(x \mid t) = -(x - t)^q_-. \tag{3.52}$$

$b^+(x \mid t), b^-(x \mid t)$ are referred to as the univariate spline basis functions for a variable x given the knot location t. The sign $+$ on the right-hand side of the parameters indicates that the two terms in Equations 3.51 and 3.52 are evaluated for positive values; that is, for $x > t$, $+(x - t)$, will be positive for all points located to the right of t, and for $x < t$, $-(x - t)$ will be positive for all the points located to the left of t, as shown in Figure 3.12.

The index q is the power to which the spline is raised in order to control the degree of smoothness of the resultant function estimate: when $q = 1$, a simple linear estimate is indicated (Sekulic and Kowalski, 1992). MARS models the true underlying function:

$$\widehat{f}(x) = a_0 + \sum_{m=1}^{M} a_m \prod_{k=1}^{K_m} B_{km} \left(x_{v(k,m)} \right), \tag{3.53}$$

where $x_1, x_2, \ldots, x_p$ are predictor variables and $x_{v(k,m)}$ labels the predictor in the kth of the mth product. K_m is a parameter that limits the order of interactions.

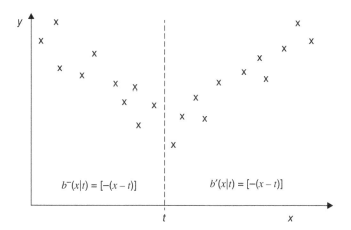

Figure 3.12 A graphical representation of a regional split applied to a univariate scatterplot (Sekulic and Kowalski (1992). Reproduced with the permission of John Wiley and Sons)

For $K_m = 1$, the resulting model will be an additive one, pairwise interactions are allowed for $K_m = 2$, and the order of interactions is arbitrary when K_m is equal to the number of compounds (n). The basis functions B_{km} are first-order truncated splines defined by Equations 3.51 and 3.52 .

The MARS algorithm proceeds as follows: a forward stepwise search for the basis function takes place with the constant basis function, the only one present initially. At each step, the split that minimized some "lack of fit" criterion from all the possible splits on each basis function is chosen. This continues until the model reaches some predetermined maximum number of basis functions, which should be about twice the number expected in the model to aid the subsequent backward stepwise deletion of the basis function.

The backward stepwise function involves removing basis functions one at a time until the "lack of fit" criterion is a minimum. In the backward stepwise deletion, the least important basis functions are eliminated one at a time. The lack of fit measure used is based on the generalized cross-validation (GCV) criterion (Craven and Wahba, 1978). The GCV is expressed as follows:

$$GCV = A \times \sum_i \left(y_i - \hat{f}(x) \right) / N, \tag{3.54}$$

with $A = [1 - C(M)/N]^{-2}$ and $C(M) = 1 + \text{trace}[B(B^- B)^{-1} B']$ being a complexity function (Friedman, 1991). The GCV criterion is the average residual error multiplied by a penalty to adjust for the variability associated with the estimation of more parameters in the model (Leblanc and Tibshirani, 1994).

The application of MARS analysis can be used to examine the relationship between rail life in terms of MGT and geometry defects.

$$\text{Rail life (MGT)} = f(\text{Geometry defects}). \tag{3.55}$$

Einbinder (2015), for example, developed a MARS relationship based on an extensive database of geometry defects.

$$MGT = A + B(BF_2) + C(BF_4) - D(BF_5), \tag{3.56}$$

where

- $BF_i = \max(0, t - x)$ where BF is the piecewise linear function for the defined variable x.
- $A = 477.37, B = 86.6031, C = 28.6544, D = 47.8978$.
- $BF_2 = \max(0, 6 - \text{warp})$; warp = number of defects
- $BF_4 = \max(0, 18 - \text{rail cant})$; rail cant = number of rail cant defects
- $BF_5 = \max(0, \text{alignment} - 0)$; alignment = number of alignment defects

Thus, if there are no track geometry defects, the calculated life, in MGT, is 1513 MGT.

This is calculated as follows:

$MGT = 477.37 + 86.6031(BF_2) + 28.6544(BF_4) - 47.8978(BF_5)$

$BF_2 = max(0, 6 - warp); warp = 0$

$BF_2 = max(0, 6 - 0) = max(0, 6); warp = 6$

$BF_4 = max(0, 18 - rail\ cant); rail\ cant = 0$

$BF_4 = max(0, 18 - 0) = max(0, 18); rail\ cant = 18$

$BF_5 = max(0, alignment - 0); alignment = 0$

$BF_5 = max(0, 0 - 0) = max(0, 0) = 0$

$MGT = 477.37 + 86.6031(6) + 28.6544(18) - 47.8978(0)$

$MGT = 477.37 + 519.62 + 515.78 - 0$

$MGT = 1513$

$$MGT = A_1 + B_1(BF_2) + C_1(BF_4) - D_1(BF_5) - E(BF_{12}), \qquad (3.57)$$

where

- $A_1 = 551.976, B_1 = 138.354, C_1 = 128.735, D_1 = 31.1105, E = 12.6995$
- $BF_2 = max(0, 5 - rail\ cant); rail\ cant = $ number of rail cant defects
- $BF_4 = max(0, 2 - warp); warp = $ number of defects
- $BF_5 = max(0, loaded\ gage - 0);$ loaded gage = number of loaded gage defects
- $BF_{12} = max(0, rail\ cant - 8); rail\ cant = $ number of rail cant defects

3.12 Deep Learning

3.12.1 General

The basic neural network algorithm and architecture is not discussed in this book, since there is a number of publications in the topic.

Deep learning refers to ML techniques that are supervised and/or unsupervised to learn automatically hierarchical representation on deep learning architectures (LeCun et al., 2015) and is more generally used to describe the learning techniques of deep models. Deep models are generally any models that involve multiple levels of computation achieved by repeated applications of nonlinear transformation. The deep learning algorithms use multiple-layer architectures or deep architecture to extract inherent features in the data from the lowest (basic level) to the highest level, thereby discovering a massive amount of structure in the data (Lv et al., 2014). These are therefore learning methods with multiple levels of representation. For example, in classification, higher layers of representation amplify aspects of the input that are relevant for discrimination and suppress irrelevant features (LeCun et al., 2015). Adding layers does not necessarily lead to better solutions. This is demonstrated in traditional neural networks, where there are issues, such as gradient descent becoming stuck in the local minima (Arnold et al., 2011).

Table 3.2 Sample of deep learning application in railway track engineering.

Authors	Application	Comments
Bai et al. (2014)	Prediction of railway track irregularities	Classification learning, tree-augmented naïve Bayes, receiver operating characteristic (ROC) curve
Berggren (2010)	Combined analysis of condition data for efficient track maintenance	Machine learning (pattern recognition)
Gibert et al. (2015)	Material classification and semantic segmentation of railway track images	Deep convolutional neural networks
Li et al. (2006)	Relating track geometry to vehicle performance	Neural networks
Xie (2014)	Learning features from high-speed train vibration signals	Deep belief networks
Yang and Létourneau (2005)	Train wheel failure prediction	Machine learning
Zhao et al. (2013)	Fault diagnosis for tuning area of jointless track circuits	Back propagation neural network

Mamoshina et al. (2016) discussed the challenges and limitations of deep learning systems compared with traditional ML methods; these include:

- The "black box" problem. The learning involves simple associations and co-occurrences.
- Overfitting and the need of large training data sets are needed.
- Selection. The selection of an appropriate deep learning system (especially the architecture) can be an issue.

A general overview of the following well-established deep architectures will be discussed in the section of this book: (a) deep belief networks, (b) convolutional neural networks, and (c) auto-encoders (Table 3.2).

Soukup and Huber-Mörk (2014) presented extensive applications of convolutional neural networks (CNN) to detect rail track defects from images. The analysis is based on laboratory tests. Figure 3.13(a) and (b) depict image patches with and without defects. The authors developed CNN with two convolutional layers with 6 and 12 output maps. A filter size of 5×5 was chosen. The pooling layers subsequently to the convolutional layers both accomplish 2×2 max-pooling and downsampling by a factor of 2 (Soukup and Huber-Mork, 2014). Figure 3.14 uses the CNN architecture. The CNN was framed in an unsupervised manner. The classification can be extended to rail structure.

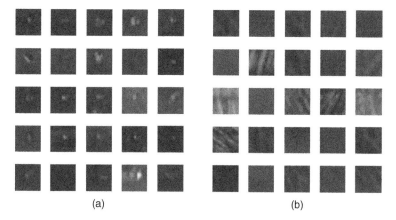

Figure 3.13 (a) Examples for surface defects and (b) non-defective samples (Soukup and Huber-Mörk, 2014). Reproduced with the permission of Springer

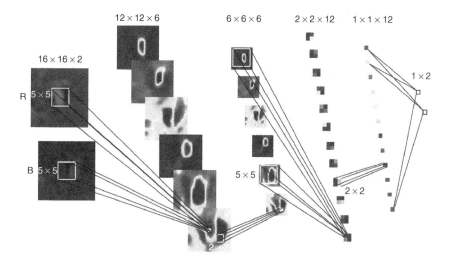

Figure 3.14 CNN architecture for surface defect detection: two convolutional and pooling layers and a final fully connected layer (Soukup and Huber-Mörk, 2014). Reproduced with the permission of Springer

3.12.2 Deep Belief Networks

3.12.2.1 Restricted Boltzmann Machines (RBM)

A restricted Boltzmann machine (RBM) is composed of an input layer and a hidden layer, which are mutually connected with an array of connection weights between the inputs and hidden neurons but no connections between neurons of the same layer. It is a bipartite graph. RBM is also a particular type of energy model (Figure 3.15).

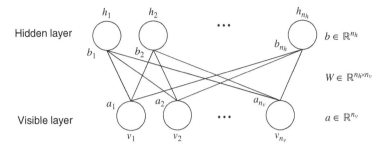

Figure 3.15 RBM structure

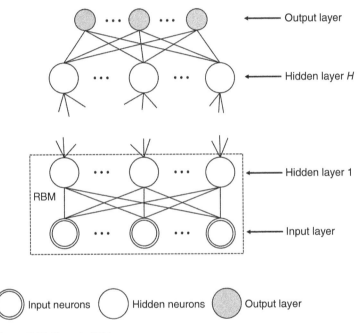

Figure 3.16 Generic DBN

3.12.2.2 Deep Belief Nets (DBN)

A deep belief net is a multilayered deep learning structure constituted by a sequence of superimposed RBM. The RBMs are probabilistic generative models that layer joint probability distribution of observed training data without using the data label (Zhao et al., 2015). A single RBM is tacked on top of the previous RBM, taking the output of previous RBM as the input after parameters of each RBM are learned properly.

Figure 3.16 shows a generic DBN. The training procedure of DBNs involves (a) pre-training and (b) fine tuning. In the pre-training phase, an unsupervised learning procedure is learned from the bottom up.

The DBN is trained by stacking RBMs layer by layer greedily during the pre-training phase to find a great parameter space. During the pre-training phase of DBN, each layer is trained as an RBM.

After the pre-training, a supervised learning algorithm is used to search the optimum space (Xu et al., 2015). The expansion of unsupervised layer-wise pre-training is the most critical improvement in deep learning algorithms. The DBN use a greedy and efficient layer-by-layer approach to learn the weights in each hidden layer. The DBN is a generative model. This implies that the models provide a joint probability distribution over observable data and labels, allowing the estimation of P(observation/label) as well as P(label/observation), while the discriminative models are limited to the latter (Arel et al., 2010). Figure 3.17 shows the training process.

3.12.3 Convolutional Neural Networks (CNN)

CNN are a family of multilayer neural networks designed for use on two-dimensional data, such as images and videos. They are composed of many layers of hierarchy with some layers for feature representation (or feature maps) and others as types of traditional neural networks (Figure 3.18). They initiate with two altering types of layers called (a) convolutional and (b) subsampling layers. The convolution layer performs the convolution operations with several filter maps of equal size. The subsampling layers reduce the sizes of proceeding layers by averaging pixels within a small neighborhood (Chen and Lin, 2014). In general, there are three types of layers defining CNN: (a) the convolutional layer, (b) the max-pooling layer, and (c) a fully connected layer.

a) *Convolutional*. Convolutional layers consist of a rectangular grid of neurons. Each convolutional layer requires that the previous layer also be a rectangular grid of neurons. Each neuron takes inputs from a rectangular section of the previous layer; the weights for this rectangular section are the same for each neuron in the convolutional layer (Gibiansky, 2014).

b) *Max-Pooling*. After each convolutional layer, there may be a pooling layer. The pooling layer takes small rectangular blocks from the convolutional layer and subsamples them to produce a single output from each block.

c) *Fully Connected*. Finally, after several convolutional and max-pooling layers, the high-level reasoning in the neural network is done via fully connected layers. A fully connected layer takes all neurons in the previous layer (be it fully connected, pooling, or convolutional) and connects it to every single neuron it has.

3.12.4 Granular Computing (Rough Set Theory)

Granular computing offers a multi-view model to describe and process uncertain, imprecise, and incomplete information. One of the learning algorithms of granular computing is rough set theory (Zhang and Lin, 2010).

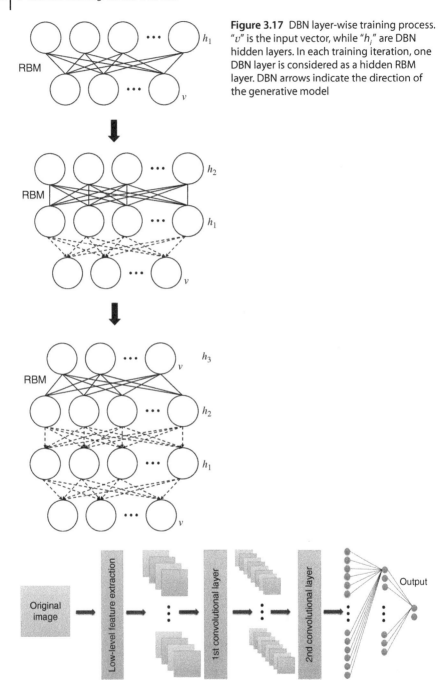

Figure 3.17 DBN layer-wise training process. "v" is the input vector, while "h_j" are DBN hidden layers. In each training iteration, one DBN layer is considered as a hidden RBM layer. DBN arrows indicate the direction of the generative model

Figure 3.18 Deep learning CNN model architecture

Rough set theory was proposed by Pawlak (1991) as a new mathematical tool for reasoning about vagueness, uncertainty, and imprecise information (Attoh-Okine, 2002). Rough set theory has been successfully applied to extracting laws from decision tables (Skowron, 1995), automated extraction of rules from clinical databases, data mining and knowledge discovery, and the infrastructure management database (Attoh-Okine, 1997). For example, a rough set was used as a data mining tool to discover minimal subsets and decision rules from the infrastructure management database. Using rough set theory, the following results (Attoh-Okine, 1997; Attoh-Okine, 2002), among others, are obtainable:

- Evaluate the importance of particular attributes in relationships between objects and decisions.
- Reduce all redundant objects and attributes and obtain minimal subsets of these attributes, which ensure a satisfactory approximation of the classification made by decisions.
- Create models of the most representative objects in particular classes of decisions.
- Represent the relationship between objects described by a reduction and decision in the form of a set of decision rules called a decision algorithm.

The mathematical framework can be accurately explained through the use of information systems. Mathematically, an information system, in the present pavement management information system, can be expressed as a 4-tuple $S = \{U, Q, V, \rho\}$, where U is a finite set of objects and Q is a finite set of attributes. $V = U_{q \in Q} V_q$ and V_q is a domain of the attributes q, and $\rho : UxQ \rightarrow V$ is a total function such that $\rho(x, q) \in V_q$ for every $q \in Q, x \in U$, called an information function.

Now consider subsets of attributes $B \subseteq Q$, where Q uniquely defines an equivalence relation:

$$\text{IND}(B) = \{(x, y) \in U2 : a(x) = a(y) \text{ for every } a \in B\}. \tag{3.58}$$

The lower approximation of $X \subseteq U$ by B is the union of all equivalence classes of $\text{IND}(B)$ that are included in X:

$$BX = \cup \{Y \in U/\text{IND}(B) : Y \subseteq X\}. \tag{3.59}$$

The upper approximation of $X \subseteq U$ by B is the union of equivalence classes that have a nonempty intersection with X, that is,

$$BX = \cup \{Y \in U/\text{IND}(B) : Y \cap X = \varnothing\}. \tag{3.60}$$

The boundary line defined as $BN_b(X) = BX - BX$ is called the boundary of X. The lower approximation consists of all elements of U that can be classified with certainty as elements of X, using knowledge B; the upper approximation is the set of all elements of U that can be possibly classified as elements of X, using attributes; the boundary set $BN_b(X)$ is the set of all elements that cannot be classified as either X or non-X by knowledge of B.

In rough set theory, decision tables are used to represent information systems. There are two basic types of decision tables that are used: a consistent decision table, where there are no conflict decision rules, and an inconsistent decision table, where there are conflicts in the decision rules.

Tables 3.3–3.6, adapted from Sadeghi and Askarinejad (2012), provide a general platform where granular computing will be applicable, especially in maintenance decision-making.

Table 3.7 is an example of a consistent pavement management database for maintenance and rehabilitation decisions. This table was reduced to Table 3.8 without the loss of any information. In Table 3.8, $U1$ contains 100 objects that are all characterized by the attribute value of { medium crack, low rutting, low raveling, medium roughness} and $D = $ {decision}.

$$\text{IND}(C) = \{\{1\}, \{2\}, \{3\}, \{4\}, \{5\}, \{6\}, \{7\}\}$$
$$\text{IND}(D) = \{\{1, 2\}, \{3, 4\}, \{5, 6, 7\}\}. \tag{3.61}$$

One can compute the reduced set of independent variables of any of the decision rules, as shown in Table 3.9. For example, the reduced family:

$$K(U6) = \{[6]a, [6]b, [6]d\}$$
$$= \{\{6, 7\}, \{4, 5, 6\}, \{5, 6, 7\}\}. \tag{3.62}$$

Table 3.3 Dominant rail group structural distresses.

Severity level	Defective component	Dominant observable distresses
Low	Rail	Bent rail, chips or dents in head, corrugation, engine burns, flaking, shelling, surface cracks.
	Joint and rail bar	Defective or missing bolts, improper type or size of bar or bolt.
Moderate	Rail	Corroded base, crushed head, end batter, side wear, vertical wear, horizontal split head, vertical split head, split web, bolt hole crack.
	Rail pad	Improper position, cracked, bent, broken, or corroded
	Joint and rail bar	Cracked bar, rail end gap >2.0 cm and <4 cm
High	Rail	Portion of rail or weld broken, horizontal split head >4 cm, vertical split head >4 cm, split web >1.5 cm, bolt hole crack >1.5 cm.
	Joint and rail bar	All bolts at joint loose, broken or missing bars, rail end gap >4 cm.

Sadeghi and Askarinejad., 2012. Reproduced with the permission of Springer.

Table 3.4 Dominant sleeper, fastening, and ballast structural distresses.

Track component	Severity level	Dominant distresses
Concrete sleeper	Low	Scaling up to 6 mm deep, cracking less than 0.80 mm in width, spalling less than 12 mm deep, efflorescence, staining
	Moderate	Scaling from 6to 25 mm deep, cracking between 0.8 and 3.20 mm, spalling between 12 and 25 mm deep
	High	Scaling more than 25 mm deep, cracking over 3.20 mm, spalling more than 25 mm deep (reinforcement steel exposed)
Timber sleeper	Low	Improperly positioned: skewed or rotated
	Moderate	Rotten, hollow, split, or impaired
	High	Broken
Ballast	Low	Dirty ballast: fine material filling voids between large aggregate particles
		Vegetation in ballast: plant life in track structure that does not interfere with track inspection
	Moderate	Vegetation in ballast that interferes with track inspection, lack or shortage of ballast in track
	High	Vegetation in ballast that interferes with train movement
Fastening	Low	Improper pattern or position
	Moderate	Loose or bent
	High	Broken or missing

Sadeghi and Askarinejad., 2012. Reproduced with the permission of Springer.

Table 3.5 Definition of track classes.

Track class	A	B	C	D
Train speed (km/h)	>160	120–160	80–120	<80

Sadeghi and Askarinejad., 2012. Reproduced with the permission of Springer.

Table 3.6 Maintenance and repair strategies.

Required maintenance and repair strategy	Structural defect density		
	Class B	Class C	Class D
No immediate action	< 0.013	< 0.033	< 0.066
Routine or preventive maintenance	0.013–0.033	0.033–0.079	0.067–0.13
Minor repair	0.034–0.066	0.080–0.13	0.14–0.30
Major repair	>0.067	>0.13	>0.30

Sadeghi and Askarinejad., 2012. Reproduced with the permission of Springer.

Table 3.7 Decision table.

U	Rail (a)	Rail and pad (b)	Joint rail and pad (c)	Decision (d)
1	1 (low)	2 (medium)	2 (medium)	1 (do nothing)
2	2 (medium)	2 (medium)	2 (medium)	1 (do nothing)
3	1 (low)	2 (medium)	2 (medium)	1 (do nothing)
4	1 (low)	2 (medium)	2 (medium)	1 (do nothing)
5	2 (medium)	2 (medium)	2 (medium)	2 (major repair)
6	1 (low)	1 (low)	2 (medium)	2 (minor repair)
7	2 (medium)	2 (medium)	2 (medium)	3 (major repair)

Table 3.8 Consistent table.

U	Rail	Rail and pad	Fastening	Joint rail and pad	Decision
1	2 (medium)	1 (low)	1 (low)	2 (medium)	2 (minor repair) 100×
2	2 (medium)	2 (medium)	1 (low)	2 (medium)	2 (minor repair) 150×
3	1 (low)	1 (low)	1 (low)	1 (low)	1 (do nothing) 120×
4	2 (medium)	2 (medium)	1 (low)	1 (low)	1 (do nothing) 50×
5	2 (medium)	2 (medium)	1 (low)	3 (high)	3 (major repair) 10×
6	3 (high)	2 (medium)	1 (low)	3 (high)	3 (major repair) 20×
7	3 (high)	3 (high)	3 (high)	3 (high)	3 (major repair) 25×

Table 3.9 Reduced set of independent variables of decision rules.

U	Rail (a)	Rail and Pad (b)	Joint rail and pad (d)	Decision (e)
1	2 (medium)	1 (low)	2 (medium)	2 (minor repair) 100×
2	2 (medium)	1 (low)	2 (medium)	2 (minor repair) 150×
3	1 (low)	1 (low)	1 (low)	1 (do nothing) 120×
4	2 (medium)	2 (medium)	1 (low)	1 (do nothing) 50×
5	2 (medium)	2 (medium)	3 (high)	3 (major repair) 10×
6	3 (high)	2 (medium)	3 (high)	3 (major repair) 20×
7	3 (high)	3 (high)	3 (high)	3 (major repair) 25×

3.12.5 Clustering

The main objective of clustering is to automatically group a data set such that similar data objects (samples) are within one cluster. Similar objects are grouped together based on a pre-selected measure. Generally, cluster analyses refer to a group of techniques used to determine the underlying structure within a data set. The aim is to partition similar objects into meaningful or useful clusters.

3.12.5.1 Measures of Similarity or Dissimilarity

Clustering methods reliy on a distance matrix or dissimilarity matrix (D) as input

$$D = \begin{bmatrix} 0 & d_{1,2} & d_{1,3} & \cdots & d_{1,n} \\ d_{2,1} & 0 & d_{2,3} & \cdots & d_{2,n} \\ d_{3,1} & d_{3,2} & 0 & \cdots & d_{3,n} \\ \vdots & \vdots & \vdots & \cdots & \vdots \\ d_{n,1} & d_{n,2} & d_{n,3} & \cdots & 0 \end{bmatrix}, \tag{3.63}$$

where $d_{i,j}$ is the distance value between pairs of observations. Calculate:

a) A Euclidean distance between two observations (p and q) measured by n variables

$$d_{p,q} = \sum_{i=1}^{n} \sqrt{\left(p_i - q_i\right)^2} \tag{3.64}$$

b) Square Euclidean

$$d_{p,q} = \sum_{i=1}^{n} \left(p_i - q_i\right)^2 \tag{3.65}$$

c) Manhattan distance

$$d_{p,q} = \sum_{i=1}^{n} \left|p_i - q_i\right| \tag{3.66}$$

d) Minkowski metric

$$d_{p,q} = \sqrt[\lambda]{\sum_{i=1}^{n} \left|p_i - q_i\right|^{\lambda}} \tag{3.67}$$

- λ can take any positive value
- $\lambda = 1$ (Manhattan distance)
- $\lambda = 2$ (Euclidean distance)

Similar or dissimilar measures used in clustering are based on the similarity or proximity of a pair of observations. The common distance functions used are:

a) Euclidean distance

$$d(x, y) = \sqrt{(x - y)'(x - y)} = \sqrt{\sum_{j=1}^{p} (x_j - y_j)^2}. \tag{3.68}$$

b) Minkowski metric

$$d(x, y) = \left[\sum_{j=1}^{p} |x_j - y_j| \right]^{1/r}. \tag{3.69}$$

There are two types of clustering methods: (a) hierarchical and (b) non-hierarchical.

3.12.5.2 Hierarchical Methods

Hierarchical clustering methods organize data into hierarchical structures of partitions starting from singleton clusters (each data point as its own cluster) to one cluster over the entire data.

The number of ways of partitioning a set of n items into g clusters is as follows (Rencher and Christensen, 2012):

$$N(n, g) = \frac{1}{g!} \sum_{k=1}^{g} \binom{g}{k} (-1)^{g-k} k^n. \tag{3.70}$$

Hierarchical clustering can further be divided into (a) the agglomerative algorithm and (b) the divisive method.

The agglomerative algorithm involves a sequential process. The process begins with h-clusters (individual items) and ends with a single cluster containing the entire data set. The divisive approach starts with a single cluster containing all n items and partitions clusters into two clusters at each step.

The agglomerative hierarchical procedure uses different approaches to measure distance (similarity measures). These include:

a) *Single linkage (nearest neighbor).* The distance between two points M, N is defined as the minimum distance between a point in M and a point in N.

$$D(M, N) = \min \{d(y_i, y_j)\}, \quad \text{for } y_i \text{ in } M \text{ and } y_j \text{ in } N \tag{3.71}$$

$$d(y_i, y_j) \rightarrow \text{Euclidean distance} \tag{3.72}$$

b) *Complete linkage (farthest neighbor).*

$$D(M, N) = \max \{d(y_i, y_j)\}, \quad \text{for } y_i \text{ in } M, y_j \text{ in } N \tag{3.73}$$

c) *Average linkage* The distance between clusters M and N is defined as the average of $k_M k_N$ distances between the k_M points in M and the k_N points in N.

$$D(M, N) = \frac{1}{k_M k_N} \sum_{i=1}^{k_M} \sum_{j=1}^{k_N} d\left(y_i, y_j\right) \tag{3.74}$$

At each step, two clusters are joined.

3.12.5.3 Non-Hierarchical Clustering

k-Means Clustering is the process of grouping similar objects together (Jain, 2010). The idea is to group objects so that the distances within clusters are minimized and the distances between clusters are maximized. Figure 3.19 presents the general concept of clustering (Galvan-Nunez and Attoh-Okine, 2016).

In addition, visualization of clusters may provide important information and can be used as a preprocessing step for other algorithms.

The process of clustering can be represented as shown in Figure 3.20, where the inputs can be presented as the similarity matrix, that is, the distance between the objects for the case of the hierarchical clustering approach and the distance between the objects to the centroids in the agglomerative clustering approach (i.e., k-means). There exist in the literature different metrics for determining the distance for non-categorical data points. In k-means, the most common distance metric used is the Euclidean distance. The output

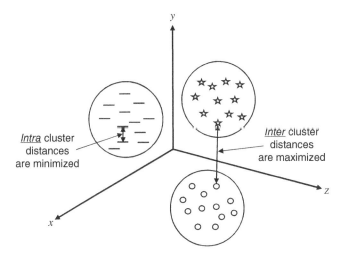

Figure 3.19 Representation of clustering (Galvan-Nunez and Attoh-Okine, 2016). Reproduced with the permission of American Society of Civil Engineers

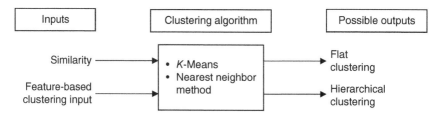

Figure 3.20 Clustering process (Galvan-Nunez and Attoh-Okine, 2016). Reproduced with the permission of American Society of Civil Engineers

obtained is a set of clusters grouped in a spherical shape (Jain, 2010). On the other hand, distance metrics have been implemented for k-means such as:

- *Mahalanobis distance*:

$$d^{(m)} = \sqrt{(x - \mu)^T \Sigma^{-1}(x - \mu)}, \tag{3.75}$$

where μ and Σ are the mean vector and covariance matrix, respectively.
- *Bregman Distance*:

$$d(x_1, x_2) = \left(\nabla \varphi(x_1) - \nabla \varphi(x_1)\right)^T (x_1 - x_2), \tag{3.76}$$

where $\varphi(x) : \Re^d \mapsto \Re$ is a strictly convex function that is twice differentiable.
- ℓ_1 *Distance*:

$$S = \sum_{i=1}^{n} \left| y_i - f(x_i) \right|, \tag{3.77}$$

where y_i is the target value and $f(x_i)$ are the estimated values.

However, this is computationally expensive; for more details, see the work of Jain (2010).

3.12.5.4   *k*-Means Algorithm

One of the most important components of a clustering algorithm is the measure of similarity used to determine how close two patterns are to one another (Niknam and Amiri, 2010). k-Means groups data vectors into a predefined number of clusters based on the Euclidean distance (see Equation 3.78) as a similarity measure:

$$D(y, c) = \sqrt{\sum_{i=1}^{n} \sum_{j=1}^{k} (y_i - c_j)^2}, \tag{3.78}$$

where

- n: number of objects or data vectors
- k: number of clusters
- y_i: ith data vector
- c_j: jth cluster centroid

The k-means process can be described as shown below:

Algorithm 3 K-means algorithm

1: Randomly initialize k centroid vectors
2: Calculate the Euclidean distance between the data vectors and the centroids using 3.78
3: **while** Stop criteria is not met **do**
4:   **for** i=1:n **do**
5:     Assign the data vector to the closest centroid vector:

$$D(y_i, c_j) = \min_{\forall l=1,\cdots,k} \{D(y_i, c_j)\} \tag{3.79}$$

6:   **end for**
7:   Update the centroid vectors:

$$c_j = \frac{1}{n_j} \sum_{\forall j=1,\cdots,k} \{D(y_i, c_j)\} \tag{3.80}$$

8: **end while**

The k-means algorithm starts by initializing the centroids randomly. Then, the distance between the data vectors and the centroids is calculated based on the Euclidean distance metric. Each data vector is assigned to one of the centroids based on the minimum distance criteria as established in Equation 3.79 . Subsequently, the jth cluster centroid is updated by summing up all the distance vectors that belong to that cluster; the cluster is then divided by n_j, that is, the number of data vectors in c_j. Equation 3.79 establishes that the data vectors are assigned iteratively to the cluster based on the minimum distance criteria. The stop criteria in k-means can be established as the maximum number of iterations when there is no significant change in the cluster centroids or when there are no changes in the cluster members after some iterations.

3.12.5.5 Expectation–Maximization (EM) Algorithms

The expectation–maximization (EM) algorithms can be seen as a type of unsupervised learning method on a mixture of models.

In railway track application, for example, let X be a data set (maybe different rail defect images) and let these images of K be assigned different objects, but it is not clear which images represent which object. Let Z be the set of latent variables that give us precisely which images represent which objects. The Z could group the rail defect (images) into the clusters. If the groupings are known, Z can be deduced, but this is not the case.

EM algorithms can be used to solve this type of clustering problem. All the images can be assigned to clusters arbitrarily. The assignments can be modified by changing which object is represented by that cluster to maximize the cluster's ability to explain the data. Updates are applied after reassigning all images to the expected most likely cluster (Zabarauskas, 2013).

The main objective of the EM algorithm is to attempt to recover the original model from the data. The algorithm is similar to k-means, and it alternates between an expectation step, corresponding to the reassignment of elements to clusters, and a maximization step corresponding to recomputation of the parameters of the model (Bordogna and Pasi, 2011).

EM Algorithms The algorithm's input are a data set X, the total number of clusters and an accepted error to converge, and the maximum number of iterations. It involves two steps:

1) Expectation step (E-step) that estimates the probability of each point belonging to each cluster
2) Maximization step (M-step) that re-estimates the parameter vector of the probability distribution of each point

The mixture models also provide a convenient approximation of the EM algorithm. Let

$$f = \sum_{k=1}^{K} \pi_k f_k \tag{3.81}$$

be a mixture of components f_k and in proportion π_k. This can be viewed as a model of K clusters, and f_k can be assumed from the same parametric family, for example, Gaussian:

$$f(x) = \sum_{k=1}^{K} \pi_k f_k(x, \mu_k, \sigma_k). \tag{3.82}$$

Let

$$p(k \mid x) = \frac{\pi_k f_k(x, \mu_k, \sigma_k)}{\sum \pi_k f_k(x)}. \tag{3.83}$$

This is the E-step.

$$\pi_k = \frac{1}{n} \sum_{i=1}^{n} p(k \mid x(i)), \tag{3.84}$$

$$\mu_k = \frac{1}{n\pi_k} \sum_{i=1}^{n} p(k \mid x(i)) x(i), \tag{3.85}$$

$$\sigma_k = \frac{1}{n\pi_k} \sum_{i-1}^{n} p(k \mid x(i)) \left(x(i) - \mu_k\right)^2. \tag{3.86}$$

This is the M-step.

The estimate proceeds in steps. The combination of E- and M-steps is guaranteed to reduce errors.

3.13   Data Stream Processing

Most of the data collection and monitoring of railway track geometry and rail defects are data intensive and involve transit data streams. The continuous arrival of data in multiple rapid, time-varying, in some cases unpredictable, and large streams requires a specific algorithm to analyze and store the data for maintenance and safety decisions. Therefore, with streaming data, there is a need for near-real-time processing, monitoring and alerting, updating, etc. One example is how "data streaming" from the grinding process can be used effectively to determine an appropriate profile.

The goal of the analysis is to convert the data stream to a compact structure that can allow for processing and queries such as Katsov (2016)

- How many distinct elements are in the monitoring data, that is, the cardinality of the data sets?
- What are the most frequent elements?
- What are the frequencies of the most frequent elements?
- How many elements belong to a specific range?

3.13.1   Methods and Analysis

There are extensive mathematical analysis-based probabilistic data structures that have been developed to analyze data streams. Some of the techniques used include (a) hash tables and (b) Bloom filters (El-Metwally et al., 2014).

- *Hash tables*: The hash table works as follows: Given a collection of data, each data entry x is stored as a record in an array. The location of the stored data is computed using a hashing function $h(x)$ that assigns each data entry to an unique integer that stands for a key that has a particular location in the array. The hashing function maps similar data entries to the same index in the hash table (Figure 3.21).
- *Bloom filters*: This is an extension of the ideas behind hash tables. It involves multiple hashes indices into a bit vector. Figure 3.22 shows an example of Bloom filter. The array bits are initially a set of zeros. The elements are then hashed several times using multiple hashing functions to obtain different hash values. The hash values should lie in a range between 0 and the size of the bloom filter.

The Bloom filter has the following probabilistic expression:

$$P(\text{false position}) = \left(1 - \left[1 - \frac{1}{m}\right]^{kn}\right)^{k} = \left(1 - e^{kn/m}\right)^{k}, \qquad (3.87)$$

where m is a bit vector, k is the number of hashes, and n is the number of elements. It has no false negatives and can be smaller and faster than hash table.

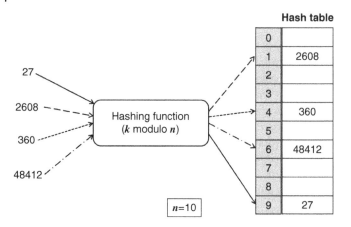

Figure 3.21 Example of a hash table (El-Metwally et al., 2014). Repoduced with the permission of Springer

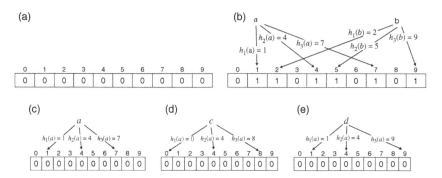

Figure 3.22 Example of a Bloom filter (El-Metwally et al., 2014). Repoduced with the permission of Springer

A linear counter that is an extension of the hash table. A single hash indexes into a bit vector. The maximum likelihood cardinality estimate is given by

$$\hat{n} = m \ln \left(\frac{\text{\# zero entries}}{m} \right).$$ (3.88)

The properties of the linear counter include

$$m > \frac{e^t - t - 1}{(\epsilon t)^2},$$ (3.89)

m bits limit the standard error to ϵ, $t = n/m$, where n is the true cardinality. It is much smaller than a hash table.

3.13.2 LogLog Counting

This well-designed hash function can transform any data set into a uniformly distributed one. The implementation requires the following parameters to be determined (Katsov, 2016):

- H – sufficient length of the hash function (in bits).
- k – number of bits that determine a bucket. 2^k is the number of buckets.
- etype – type of the estimator (e.g., byte), that is, how many bits are required for each estimator.

Table 3.10 presents the equations needed LogLog Counter algorithm.

3.13.3 Count–Min Sketch

This is similar to linear counting. The count–min sketch is simply a two-dimensional array ($d \times w$) of integer counters (Figure 3.23, Table 3.11).

Table 3.10 Equations to determine numerical parameters of the LogLog Counter.

Equation	Definition
$\hat{n} = \alpha_m \cdot m \cdot 2^{1/m \sum_j \text{estimators}(j)}$	$\hat{n}$ – cardinality estimation
	m – number of buckets (estimators)
$\alpha_m = \Gamma\left(\dfrac{-1}{m}\right) \dfrac{1-2^{1/m}}{\ln 2}^{m>64} \approx 0.39701$	α_m – estimation factor, close to 0.39701 for $m > 64$, that is, for most practical problems
$\varepsilon \approx \dfrac{1.30}{\sqrt{m}}$	Dependency between the standard error of the estimation and the number of buckets (estimators)
	ε – standard error of the estimation
	m – number of buckets (estimators)
$H = \log_2 m + [\log_2(n/m) + 3]$	A practical formula for length of the hash function
	m – number of buckets (estimators)
	n – maximum cardinality (i.e., capacity)
etype $\Leftarrow [\log_2[\log_2(n/m) + 3]]$	The number of bits in etype is determined by the maximal possible rank. The rank is limited by H, so the length of etype is a log of H (except the part that is used for bucket ID computation).

Courtesy: Ilya Katsov.

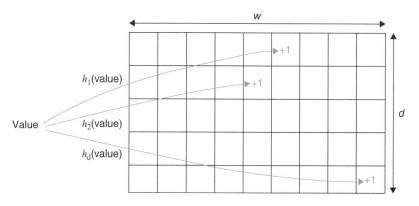

Figure 3.23 Count–min sketch idea

Table 3.11 Dependency between the sketch size and accuracy.

Equation	Definition
Estimation error $\varepsilon \leq 2n/w$ with probability $\delta = 1 - (1/2)^d$	n – total count of registered events w – sketch width d – sketch height (a.k.a. depth)

Courtesy: Katsov, 2016.

Table 3.12 Streaming techniques.

Technique	Purpose
Bloom filter	Membership query
Linear counting	Cardinality estimation
LogLog counting	
Count–min sketch	Frequency estimation
Count–mean–min sketch	

Table 3.12 shows streaming techniques.

Table 3.13 shows some applications of machine learning techniques in railway track engineering.

Instrumented wheelsets (IWS) provide accurate measurements of wheel/rail contact forces, and the data is processed in real time. The information calculated includes

a) Wheel side L/V ratio, which addresses wheel climbing rail
b) Thick side L/V ratio, which addresses rail rollover
c) Axle-sum L/V ratio, which addresses climbing rail and track lateral shift
d) Wheel unloading, which involves wheel climbing rail

Wahba (2012) presented IWS applications and highlighted that continuous track monitoring is very important. This will provide benefits such as

a) Optimization of wheel/rail interface
b) Optimization and prioritization track maintenance
c) Decreased safety risk

Figure 3.24 allows the application of streaming data to be analyzed appropriately. Figure 3.25 shows IWS application.

Data streaming methods can be applied to condition monitoring and predictive maintenance of train bearings based on sensor data output. The CBM approach is based on constantly monitoring the conditions of an asset

Table 3.13 Application of machine learning techniques.

Authors (year)	Application	Source of data	Research aim	Machine learning technique	Comments
Andrade and Teixeira (2013)	Modeling of rail track geometry degradation	Main Portuguese railway line Lisbon-Oporto	Predict rail track geometry degradation and thus guide planning maintenance and renewal actions.	Hierarchical Bayesian models, Markov chain Monte Carlo	
Andrews (2012)	Railway track asset management	UK Network Rail system	Investigate alternative strategies to effectively manage railway track assets	Petri nets with Monte Carlo Simulation	
Bai et al. (2015)	Prediction of railway track irregularities	Chinese railway system	Evaluate the deterioration of track maintenance units	Markov-based models	
Bai et al. (2016)	Prediction of railway track irregularities	Chinese railway system	Identify possible underlying patterns or rules for predicting track irregularities based on the characteristic deterioration of track maintenance units	Classification learning, tree-augmented naïve Bayes, Receiver operating characteristic (ROC) curve)	
Berggren (2010)	Combined analysis of condition data for efficient track maintenance	Swedish railway	Eliciting new information form measured condition data originating from track geometry quality, dynamic stiffness, and ground penetrating radar	Machine learning (Pattern recognition)	
Bouillaut et al. (2008)	Rail maintenance strategy model aimed at the prevention of broken rails	Régie Autonome des Transports Parisiens (RATP), France	Identify precisely the impact of rail flaws on safety and availability of the railway system	Dynamic Bayesian networks	

(Continued)

Table 3.13 (Continued)

Authors (year)	Application	Source of data	Research aim	Machine learning technique	Comments
Fumeo et al. (2015)	Condition-based maintenance in railway transportation systems	FEMTO-ST Institute, France	Addresses with the problem of condition-based maintenance (CBM) applied to the condition monitoring and predictive maintenance of train axle bearings based on sensors data collection, with the purpose of maximizing their remaining useful life (RUL)	Big data streaming analysis, online support vector regression models	
Gibert et al. (2015)	Material classification and semantic segmentation of railway track images	Amtrak; Federal Railroad Administration	Extraction of accurate information from visual track inspection images	Deep convolutional neural networks	
He et al. (2011)	Modal parameter identification of railway bridges	Nanjing Yangtze River Bridge	Identify modal parameters from monitoring vibrational data	Empirical mode decomposition	
Ho et al. (2010)	Rail structure analysis	Taoyuan Taiwan Railway; Simulation dataset	Explore the feasibility and applicability of using EMD and HHT on the analysis of track structure system through laboratory testing and vibration testing	Empirical mode decomposition	
Lam et al. (2014)	Identification of railway ballast damage under a concrete sleeper	Mass Transit Railway (MTR) Corporation, Hong Kong	Address the problem of detecting railway ballast damage under a concrete sleeper	Bayesian analysis (Bayesian probabilistic approach)	
Li et al. (2006)	Relating track geometry to vehicle performance	Transportation Technology Center, Inc., Association of American Railroads	Recognizing complex patterns and nonlinear relationship between track geometry and vehicle response	Neural networks	

Reference	Application	Data	Description	Method
Li et al. (2013)	Alarm prediction in large-scale sensor networks	US Class I railroad	Predict alarms associated with catastrophic equipment failures several days ahead of time	Support vector machines
Li et al. (2016)	Railway wheel flat detection	Simulation data sets	Examine the axle box vibration response caused by wheel flats, considering the influence of both track irregularity and vehicle running speed on diagnosis results	Empirical mode decomposition
Oukhellou et al. (2006)	Railway infrastructure system diagnosis (track circuit diagnosis)	French high-speed line (LGV), Simulation data set	Detect its working state from one measurement signal which can be viewed as a superposition of several oscillations and periodic patterns called intrinsic mode functions (IMFs)	Empirical mode decomposition
Quiroga and Schnieder (2012)	Railway track geometry deterioration and restoration	French railway operator SNCF	Modeling and simulation of the track geometry aging and restoration process	Monte Carlo simulation
Prescott and Andrews (2015)	Railway track asset management	UK rail network	Investigate asset management strategies applied to railway track sections	Markov model
Sinha and Feroz (2016)	Obstacle detection on railway tracks	High MARNHAM Rail Innovation and Development Centre, UK	Deployment of (Micro-Electro mechanical systems) MEMS sensors in railway track monitoring in order to detect obstacles like rock and timber dropping along the track.	Bayesian analysis
Sun and Zhao (2013)	Fault diagnosis in railway track circuits	Simulation data set	Ensure a high level of safety and availability of track circuit	Support vector machines

(Continued)

Table 3.13 (Continued)

Authors (year)	Application	Source of data	Research aim	Machine learning technique	Comments
Sun et al. (2012)	Fault diagnosis for railway track circuits trimming capacitors	Simulation data set	Ensure required dependability and availability levels of track circuit	Empirical mode decomposition	
Shu et al. (2014)	On-site detection of the stress-free temperature of a continuous welded rail	Hohhot Railways Bureau, China	Appropriate selection of stress-free temperature in each region when laying continuous welded rail	Support vector machines	
Tang et al. (2015)	Prediction of high-speed railway settlement	Chinese high-speed railway line	Stable accurate prediction of high-speed railway settlement	Wavelet neural network	
Tsai et al. (2014)	Railway track inspection	Taiwan High-Speed Rail Corporation (THSRC)	Detect the rapid development of defects on railway systems	Empirical mode decomposition	
Vileiniskis et al. (2015)	Fault detection for railway point systems (RPS)	UK Network Rail	Early detection of the changes in the measurement of the current drawn by the motor of the point operating equipment (POE) of an RPS, which can be used to warn about a possible failure in the system	Support vector machines	
Wang et al. (2015)	Detecting the loss of rail fastening nuts	Short abandoned rail line in Hangzhou, China	Inspection method to reduce maintenance costs and improve maintenance efficiency	Support vector machines	
Wellalage et al. (2013)	Predicting future conditions of railway bridge elements	Australian railway bridges	Overcome invalid future condition prediction in existing methods	Markov chain Monte Carlo	

Reference	Title	Data set	Objective	Method
Xie (2014)	Learning features from high-speed train vibration signals	Anonymous real data set and simulation data sets	Automatically extract high-level features from HST vibration signals and recognize the faults.	Deep belief networks
Yang and Létourneau (2005)	Train Wheel Failure Prediction	Canadian railroad	Optimize maintenance and operation of trains through prognostics of wheel failures	Machine learning
Zhao et al. (2013)	Fault diagnosis for tuning area of jointless track circuits	Nanning Railway Bureau, China	Overcome the disadvantages of the current detection methods in aspects such as detection cost and timeliness	Back propagation neural network
Zhao et al. (2014)	High-speed rail fault diagnosis	Simulation data set	Recognize fault patterns accurately and effectively	Empirical mode decomposition
Zilko et al. (2016)	Modeling of railway disruption length	Dutch railway network	To assist the Dutch Operational Control Center Rail (OCCR) during disruptions	Copula Bayesian networks
Zhu et al. (2015)	Railway line state Detection signal processing (signal processing of track state detection)	Simulation data set	Accuracy of measurement, which is regularly influenced by the noise in the output signal of inertial measurement unit in the track state detecting process	Empirical mode decomposition

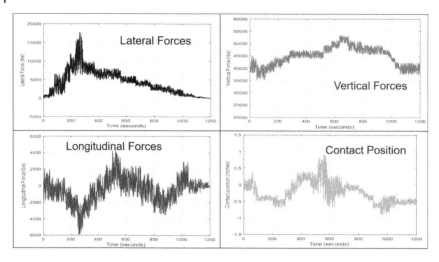

Figure 3.24 IWS sample signal

Examples of IWS applications

| In-service monitoring | Continuous track monitoring (streaming) |

Figure 3.25 IWS application

and performing maintenance on demand if any potential asset degradation is detected. In the SDA for CBM algorithms, there are two challenges (Fumeo et al., 2015):

a) Training and selecting accurate degradation models (which can be based on physical laws or data driven)
b) Deploying such a strategy in order to optimize the trade-off between computational requirements and accuracy

Fumeo et al. (2015) used the online support vector regression (O-SVR) algorithm to estimate remaining useful life (RUL) in service axle bearings. The online algorithm updates trained SVR first whenever there was a new sample (x_{n+1}, y_{n+1}) added to training set D_n.

3.13.3.1 Online Support Regression
The conventional support vector regression (SVR) searches a regressor in the form $f(x) = w^T \phi(x) + b$, which is a linear regressor in the space defined by the

nonlinear function $\phi : \mathbb{R}^d \rightarrow \mathbb{R}^d$ where usually $D \gg d$ (Fumeo et al., 2015),

$$\min_{w,b,\xi^+,\xi^-} \frac{1}{2}\|w\|^2 + C \sum_{i=1}^{n} \left(\xi_i^+ + \xi_i^-\right),$$

$$\text{s.t.} \begin{cases} y_i - w^T \phi(x_i) - b \leq \epsilon + \xi_i^+, \ \forall i \in \{1,\dots,n\} \\ w^T \phi(x_i) + b - y_i \leq \epsilon + \xi_i^-, \ \forall i \in \{1,\dots,n\} \,, \\ \xi_i^+, \xi_i^- \geq 0 \end{cases} \tag{3.90}$$

where $\epsilon \geq 0$ is the maximum deviation allowed during the training and $C \geq 0$ is the associated penalty for excess deviation during the training.

$$\max \frac{1}{2} \sum_{i,j=1}^{m} \left(\alpha_i - \alpha_i^*\right)\left(\alpha_j - \alpha_j^*\right)\langle x_i, x_j \rangle - \sum_{i=1}^{m}\left(\alpha_i + \alpha_i^*\right) + \sum_{i=1}^{m} y_i \left(\alpha_i - \alpha_i^*\right),$$

$$\text{s.t.} \sum_{i=1}^{m}(\alpha_i - \alpha_i^*) = 0, \ 0 \leq \alpha_i, \alpha_i^* \leq 0. \tag{3.91}$$

3.14 Remarks

Table 3.13 shows the application of ML techniques in railway track engineering. Most of these applications have been implemented. The next issue is how similar problems can be solved in the presence of large amounts of data and streaming.

Another major improvement can be the application of hybrid methods, combinations of two or more ML techniques. Streaming methods can be used to analyze lateral forces, vertical forces, longitudinal forces, and contact position.

References

E. Alpaydın. Machine learning. *Wiley Interdisciplinary Reviews: Computational Statistics*, **3**(3):195–203, 2011. doi: 10.1002/wics.166.

A. R. Andrade and P. F. Teixeira. Hierarchical Bayesian modeling of rail track geometry degradation. *Proceedings of the Institution of Mechanical Engineers, Part F: Journal of Rail and Rapid Transit*, **227**(4):364–375, 2013. doi: 10.1177/0954409713486619.

J. Andrews. A modeling approach to railway track asset management. *Proceedings of the Institution of Mechanical Engineers, Part F: Journal of Rail and Rapid Transit*, **227**(1):56–73, 2012. doi: 10.1177/0954409712452235.

I. Arel, D. C. Rose, and T. P. Karnowski. Deep machine learning – a new frontier in artificial intelligence research. *IEEE Computational Intelligence Magazine*, **5**(4):13–18, 2010. doi: 10.1109/MCI.2010.938364.

L. Arnold, S. Rebecchi, S. Chevallier, and H. Paugam-Moisy. An introduction to deep learning. In *European Symposium on Artificial Neural Networks, Computational Intelligence and Machine Learning*, pages 477–488. ESANN 2011 proceedings, April 2011. doi: 10.1109/ISSPA.2012.6310529.

N. O. Attoh-Okine. Rough set application to data mining principles in pavement management database. *Journal of Computing in Civil Engineering*, **11**(4):231–237, 1997. doi: 10.1061/(ASCE)0887-3801(1997)11:4(231).

N. O. Attoh-Okine. Combining use of rough set and artificial neural networks in doweled-pavement-performance modeling – a hybrid approach. *Journal of Transportation Engineering*, **128**(3):270–275, 2002. doi: 10.1061/(ASCE)0733-947X(2002)128:3(270).

L. Bai, R. Liu, Q. Sun, F. Wang, and F. Wang. Classification-learning-based framework for predicting railway track irregularities. *Proceedings of the Institution of Mechanical Engineers, Part F: Journal of Rail and Rapid Transit*, **230**(2):598–610, 2014. doi: 10.1177/0954409714552818.

L. Bai, R. Liu, Q. Sun, F. Wang, and P. Xu. Markov-based model for the prediction of railway track irregularities. *Proceedings of the Institution of Mechanical Engineers, Part F: Journal of Rail and Rapid Transit*, **229**(2):150–159, 2015. doi: 10.1177/0954409713503460.

L. Bai, R. Liu, Q. Sun, F. Wang, and F. Wang. Classification-learning-based framework for predicting railway track irregularities. *Proceedings of the Institution of Mechanical Engineers, Part F: Journal of Rail and Rapid Transit*, **230**(2):598–610, 2016. doi: 10.1177/0954409714552818.

E. G. Berggren. Efficient track maintenance: methodology for combined analysis of condition data. *Proceedings of the Institution of Mechanical Engineers, Part F: Journal of Rail and Rapid Transit*, **224**(5):353–360, 2010. doi: 10.1243/09544097JRRT354.

G. Bordogna and G. Pasi. Soft clustering for information retrieval applications. *Wiley Interdisciplinary Reviews: Data Mining and Knowledge Discovery*, **1**(2):138–146, 2011. doi: 10.1002/widm.3.

L. Bouillaut, O. Francois, P. Leray, P. Aknin, and S. Dubois. Dynamic Bayesian networks modeling maintenance strategies: prevention of broken rails. In *Proceedings of 8th World Congress on Railway Research {WCCR'08}*, 2008.

G. Camps-Valls and L. Bruzzone. *Kernel Methods for Remote Sensing Data Analysis*, 2009. http://onlinelibrary.wiley.com/doi/10.1002/9780470748992.fmatter/summary.

D.-S. Cao, Q.-S. Xu, Y.-Z. Liang, L.-X. Zhang, and H.-D. Li. The boosting: a new idea of building models. *Chemometrics and Intelligent Laboratory Systems*, **100**(1):1–11, 2010. doi: 10.1016/j.chemolab.2009.09.002.

X.-W. Chen and X. Lin. Big data deep learning: challenges and perspectives. *IEEE Access*, **2**:514–525, 2014. doi: 10.1109/ACCESS.2014.2325029.

P. Craven and G. Wahba. Smoothing noisy data with spline functions. *Numerische Mathematik*, **31**(4):377–403, 1978. doi: 10.1007/BF01404567.

D. Einbinder. *The development of rail defects due to the presence of geometry defects in class 1 railroads*. PhD thesis, Master Thesis. University of Delaware, 2015. http://dspace.udel.edu/bitstream/handle/19716/17394/2015_EinbinderDaniel_MCE.pdf?sequence=1&isAllowed=y.

S. El-Metwally, O. M. Ouda, and M. Helmy. Algorithms and data structures in next-generation sequencing. In *Next Generation Sequencing Technologies and Challenges in Sequence Assembly*, volume **7**, Chapter 2, *SpringerBriefs in Systems Biology*, pages 15–25. Springer, New York, 2014. ISBN: 978-1-4939-0714-4.

J. H. Friedman. Multivariate adaptive regression splines. *Annals of Statistics*, **19**(1):1–67, 1991. doi: 10.1152/japplphysiol.00729.2009.

Y. Fu. Kernel methods and applications in bioinformatics. In *Handbook of Bio-/Neuroinformatics*, pages 275–285. Springer, Berlin Heidelberg, 2014. doi: 10.1007/978-3-642-30574-0_18.

G. Fu, R. Dawson, M. Khoury, and S. Bullock. Interdependent networks: vulnerability analysis and strategies to limit cascading failure. *The European Physical Journal B*, **87**(7):148, 2014. doi: 10.1140/epjb/e2014-40876-y.

E. Fumeo, L. Oneto, and D. Anguita. Condition-based maintenance in railway transportation systems based on big data streaming analysis. *Procedia Computer Science*, **53**:437–446, 2015. doi: 10.1016/j.procs.2015.07.321.

M. Galar, A. Fernandez, E. Barrenechea, H. Bustince, and F. Herrera. A review on ensembles for the class imbalance problem: bagging, boosting, and hybrid-based approaches. *IEEE Transactions on Systems, Man, and Cybernetics, Part C (Applications and Reviews)*, **42**(4):463–484, 2012. doi: 10.1109/TSMCC.2011.2161285.

S. Galvan-Nunez and N. Attoh-Okine. Hybrid particle swarm optimization and K-means analysis for bridge clustering based on national bridge inventory data. *ASCE-ASME Journal of Risk and Uncertainty in Engineering Systems, Part A: Civil Engineering*, 2016. doi: 10.1061/AJRUA6.0000864.

X. Gibert, V. M. Patel, and R. Chellappa. Deep Multi-Task Learning for Railway Track Inspection, 2015. http://arxiv.org/abs/1509.05267.

A. Gibiansky. Convolutional Neural Networks, 2014. http://andrew.gibiansky.com/blog/machine-learning/convolutional-neural-networks/.

D. Gutierrez. Ask a Data Scientist: The Bias vs. Variance Tradeoff, 2014. http://insidebigdata.com/2014/10/22/ask-data-scientist-bias-vs-variance-tradeoff/.

X. H. He, X. G. Hua, Z. Q. Chen, and F. L. Huang. EMD-based random decrement technique for modal parameter identification of an existing railway bridge. *Engineering Structures*, **33**(4):1348–1356, 2011. doi: 10.1016/j.engstruct.2011.01.012.

H. Ho, P. Chen, D. T. Chang, and C. Tseng. Rail structure analysis by empirical mode decomposition and Hilbert Huang transform. *Tamkang Journal of Science and Engineering*, **13**(3):267–279, 2010.

A. K. Jain. Data clustering: 50 years beyond K-means. *Pattern Recognition Letters*, **31**(8):651–666, 2010. doi: 10.1016/j.patrec.2009.09.011.

I. Katsov. Probabilistic Data Structures for Web Analytics and Data Mining, 2016. https://highlyscalable.wordpress.com/2012/05/01/probabilistic-structures-web-analytics-data-mining/.

S. B. Kotsiantis. Bagging and boosting variants for handling classifications problems: a survey. *The Knowledge Engineering Review*, **29**(1):78–100, 2013. doi: 10.1017/S0269888913000313.

H. F. F. Lam, Q. Hu, and M. T. T. Wong. The Bayesian methodology for the detection of railway ballast damage under a concrete sleeper. *Engineering Structures*, **81**:289–301, 2014. doi: 10.1016/j.engstruct.2014.08.035.

M. Leblanc and R. Tibshirani. Adaptive principal surfaces. *Journal of the American Statistical Association*, **89**(425):53–64, 1994. doi: 10.1080/01621459.1994 .10476445.

Y. LeCun, Y. Bengio, and G. Hinton. Deep learning. *Nature*, **521**:436–444, 2015. doi: 10.1038/nmeth.3707.

B. K. Li, B. He, Z. Y. Tian, Y. Z. Chen, and Y. Xue. Modeling, predicting and virtual screening of selective inhibitors of MMP-3 and MMP-9 over MMP-1 using random forest classification. *Chemometrics and Intelligent Laboratory Systems*, **147**:30–40, 2015. doi: 10.1016/j.chemolab.2015.07.014.

Y. Li, J. Liu, and Y. Wang. Railway wheel flat detection based on improved empirical mode decomposition. *Shock and Vibration*, **2016**:1–14, 2016.

D. Li, A. Meddah, K. Hass, and S. Kalay. Relating track geometry to vehicle performance using neural network approach. *Proceedings of the Institution of Mechanical Engineers, Part F: Journal of Rail and Rapid Transit*, **220**(3):273–281, 2006. doi: 10.1243/09544097JRRT39.

H. Li, B. Qian, D. Parikh, and A. Hampapur. Alarm prediction in large-scale sensor networks – a case study in railroad. In *2013 IEEE International Conference on Big Data*, pages 7–14. IEEE, 2013. ISBN: 978-1-4799-1293-3.

H. Lodhi. Computational Biology Perspective: Kernel Methods and Deep Learning, 2012. ISSN: 19395108.

Y. Lv, Y. Duan, W. Kang, Z. Li, and F. Y. Wang. Traffic flow prediction with big data: a deep learning approach. *IEEE Transactions on Intelligent Transportation Systems*, **16**(2):865–873, 2014. doi: 10.1109/TITS.2014.2345663.

X. Ma, H. Yu, Y. Wang, and Y. Wang. Large-scale transportation network congestion evolution prediction using deep learning theory. *PLoS One*, **10**(3):1–17, 2015. doi: 10.1371/journal.pone.0119044.

P. Mamoshina, A. Vieira, E. Putin, and A. Zhavoronkov. Applications of deep learning in biomedicine. *Molecular Pharmaceutics*, **13**:1445–1454, 2016. doi: 10.1021/acs.molpharmaceut.5b00982.

T. Niknam and B. Amiri. An efficient hybrid approach based on PSO, ACO and k-means for cluster analysis. *Appl. Soft Comput.*, **10**(1):183–197, 2010.

L. Oukhellou, P. Aknin, and E. Delechelle. Railway infrastructure system diagnosis using empirical mode decomposition and Hilbert transform. In *2006 IEEE International Conference on Acoustics Speed and Signal Processing Proceedings*, volume 3(1), III–1164–III–1167, 2006. doi: 10.1109/ICASSP.2006.1660866.

Z. Pawlak. *Rough Sets: Theoretical Aspects of Reasoning About Data*. Kluwer Academic Publishers, 1991. ISBN: 0792314727.

D. Prescott and J. Andrews. Investigating railway track asset management using a Markov analysis. *Proceedings of the Institution of Mechanical Engineers Part F-Journal of Rail and Rapid Transit*, **229**(4):402–416, 2015. doi: 10.1177/0954409713511965.

L. M. Quiroga and E. Schnieder. Monte Carlo simulation of railway track geometry deterioration and restoration. *Proceedings of the Institution of Mechanical Engineers, Part O: Journal of Risk and Reliability*, **226**(3):274–282, 2012. doi: 10.1177/1748006X11418422.

A. C. Rencher and W. F. Christensen. Cluster analysis. In *Methods of Multivariate Analysis*, pages 501–554. John Wiley & Sons, Inc., 3rd edition, 2012. doi: 10.1002/9781118391686.ch15.

J. Sadeghi and H. Askarinejad. Application of neural networks in evaluation of railway track quality condition. *Journal of Mechanical Science and Technology*, **26** (1):113–122, 2012. doi: 10.1007/s12206-011-1016-5.

B. Scholkopf, K. Tsuda, and J.-P. Vert. *Kernel Methods in Computational Biology*. MIT Press, 2004. ISBN: 9780262195096.

S. Sekulic and B. R. Kowalski. MARS: a tutorial. *Journal of Chemometrics*, **6**(4):199–216, 1992. doi: 10.1002/cem.1180060405.

P. Sephton. Forecasting recessions: can we do better on MARS? *Federal Reserve Bank of St. Louis Review*, **83**(2):39–49, 2001. http://ideas.repec.org/a/fip/fedlrv/y2001imarp39-49nv.83no.2.html.

D. Shu, L. Yin, J. Bu, J. Chen, and X. Qi. Application of a combined metal magnetic memory-magnetic Barkhausen noise technique for on-site detection of the stress-free temperature of a continuous welded rail. *Proceedings of the Institution of Mechanical Engineers, Part F: Journal of Rail and Rapid Transit*, **20**(3):774–783, 2014. doi: 10.1177/0954409714562874.

D. Sinha and F. Feroz. Obstacle detection on railway tracks using vibration sensors and signal filtering. *IEEE Sensors Journal*, **16**(3):642–649, 2016. doi: 10.1109/JSEN.2015.2490247.

A. Skowron. Extracting laws from decision tables: a rough set approach. *Computational Intelligence*, **11**(2):371–388, 1995. doi: 10.1111/j.1467-8640.1995.tb00039.x.

D. Soukup and R. Huber-Mörk. Convolutional neural networks for steel surface defect detection from photometric stereo images. *Advances in Visual Computing*, **8887**:668–677, 2014. doi: 10.1007/978-3-319-14249-4_64.

D. Steinberg. An alternative to neural nets: multivariate adaptive regression splines (MARS). *PC AI*, **15**(1):38–41, 2001. ISSN: 0894-0711.

S. Sun and H. Zhao. Fault diagnosis in railway track circuits using support vector machines. In *Proceedings of the 12th International Conference on Machine Learning and Applications*, volume 2, pages 345–350, 2013. doi: 10.1109/ICMLA.2013.146.

S. P. Sun, H. B. Zhao, and G. Zhou. A fault diagnosis approach for railway track circuits trimming capacitors using EMD and Teager energy operator. *WIT Transactions on the Built Environment*, **127**:167–176, 2012. doi: 10.2495/CR120151.

S. Tang, F. Li, Y. Liu, L. Lan, C. Zhou, and Q. Huang. Application of wavelet neural network model based on genetic algorithm in the prediction of high-speed railway settlement. In *Proceedings of SPIE 9808, International Conference on Intelligent Earth Observing and Applications 2015*, page 98082P, 2015. doi: 10.1117/12.2222200.

H.-C. Tsai, C.-Y. Wang, N. E. Huang, T.-W. Kuo, and W.-H. Chieng. Railway track inspection based on the vibration response to a scheduled train and the Hilbert-Huang transform. *Proceedings of the Institution of Mechanical Engineers, Part F: Journal of Rail and Rapid Transit*, **229**(7):815–829, 2014. doi: 10.1177/0954409714527930.

M. Vileiniskis, R. Remenyte-Prescott, and D. Rama. A fault detection method for railway point systems. *Proceedings of the Institution of Mechanical Engineers, Part F: Journal of Rail and Rapid Transit*, 1–14, 2015. doi: 10.1177/0954409714567487.

A. Wahba. The Use of NRC Instrumented Wheelsets in Revenue Service (Now and in the Future), 2012. http://www.wheel-rail-seminars.com/downloads.php.

L. Wang, B. Zhang, J. Wu, H. Xu, X. Chen, and W. Na. Computer vision system for detecting the loss of rail fastening nuts based on kernel two-dimensional principal component – two-dimensional principal component analysis and a support vector machine. *Proceedings of the Institution of Mechanical Engineers, Part F: Journal of Rail and Rapid Transit*, **230**(8):1842–1850, 2015. doi: 10.1177/0954409715616426.

N. K. W. Wellalage, T. Zhang, and R. Dwight. Calibrating Markov Chain – based deterioration models for predicting future conditions of railway bridge elements. *Journal of Bridge Engineering*, **20**(2):04014060, 2013. doi: 10.1061/(ASCE)BE.1943-5592.0000640.

J. Xie. Learning features from high-speed train vibration signals with deep belief networks. In *2014 International Joint Conference on Neural Networks (IJCNN)*, pages 2205–2210, 2014.

J. Xu, H. Li, and S. Zhou. An overview of deep generative models. *IETE Technical Review*, **32**(2):131–139, 2015. doi: 10.1080/02564602.2014.987328.

C. Yang and S. Létourneau. Learning to predict train wheel failures. In *Proceeding of the 11th ACM SIGKDD International Conference on Knowledge Discovery in*

Data Mining – KDD '05, page 516, ACM Press, New York, 2005. doi: 10.1145/1081870.1081929.

M. Zabarauskas. Expectation-Maximization Algorithm for Bernoulli Mixture Models (Tutorial), 2013. http://blog.manfredas.com/expectation-maximization-tutorial/.

W. J. Zhang and Y. Lin. On the principle of design of resilient systems – application to enterprise information systems. *Enterprise Information Systems*, **4**(2):99–110, 2010. doi: 10.1080/17517571003763380.

X. Zhao, X. Shi, and S. Zhang. Facial expression recognition via deep learning facial expression recognition via deep learning. *IETE Technical Review*, **32**(5):347–355, 2015. doi: 10.1080/02564602.2015.1017542.

J. Zhao, Y. Yang, T. Li, and W. Jin. Application of empirical mode decomposition and fuzzy entropy to high-speed rail fault diagnosis. In *Advances in Intelligent Systems and Computing (AISC)*, pages 93–103, 2014. doi: 10.1007/978-3-642-54924-3_9.

L.-H. Zhao, C.-L. Zhang, K.-M. Qiu, and Q.-L. Li. A fault diagnosis method for the tuning area of jointless track circuits based on a neural network. *Proceedings of the Institution of Mechanical Engineers Part F-Journal of Rail and Rapid Transit*, **227**(4):333–343, 2013. doi: 10.1177/0954409713480453.

W.-F. Zhu, H.-Z. Ma, X.-D. Chai, and S.-B. Zhen. The research of railway line state detection signal processing method based on EMD. *Open Journal of Safety Science and Technology*, **5**:63–68, 2015.

A. A. Zilko, D. Kurowicka, and R. M. P. Goverde. Modeling railway disruption lengths with copula Bayesian networks. *Transportation Research Part C: Emerging Technologies*, **68**:350–368, 2016. doi: 10.1016/j.trc.2016.04.018.

4

Basic Foundations of Big Data

4.1   Introduction

Frequent monitoring of railway track profile and geometry is very critical for planning proper and cost-effective maintenance. These inspections provide information like vertical irregularities of the rails, real altitude differences between two rail running surfaces, and other features. This data, which is collected in real time from time series, can be used to develop both maintenance and safety plans for the railway and other conditions that could cause safety problems on the tracks. This information can be used to detect and analyze imminent problems with tracks. Also, visual sensors that shoot ultrasound visuals can detect when wheels flatten and ultimately can lead to a reduction in bearing-related derailments. This is important considering the fact that a major derailment can cost as much as $40 million (Hunt et al., 2012).

Big data analytics in railway track engineering research focused on collecting, examining, and processing large multiple-source data sets (structured, unstructured, and streaming) to discover patterns and correlations and extract information from the data for maintenance and safety decisions.

There are common myths associated with big data. Jagadish (2015) noted the following:

a) Size is all that matters – Big data is not all about size. The five Vs' methodology was proposed after an initial emphasis on size. It is now clear that variety appears to be difficult; also, the volume and velocity can be challenging.
b) The central challenge with big data is that of devising new computing architecture and algorithms – This can be a little misleading. Lately, there have been growing applications like MapReduce, which keeps changing every few months. Most of the architecture is focused on volume, and in some cases velocity, but little consideration on variety and veracity.
c) Analytics is the central problem with big data – There are many steps and processes to big data. Each step has its own outcome. Figure 4.1 presents the big data analysis pipeline.

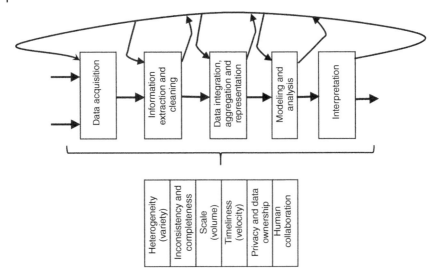

Figure 4.1 The big data analysis pipeline (Jagadish, 2015). Reproduced with the permission of Elsevier

d) Data reuse is low-hanging fruit – Original data needs to be used at the appropriate time. This implies the topic of interest should be clear in the data. Furthermore, data may not conform to the required application.

e) Data science is the same as big data – The primary difference between the two is that big data begins with data characteristics (and works up from there), while data science begins with data use (and works down from there).

f) Big data is all hype – Advanced data analysis has been around for a long time. The only changing parameters are the amount and characteristics of data being collected, which keep changing with different formats.

Fan et al. (2014) presented some salient features of big data and noted that big data create unique features that are not shared or presented by traditional data sets. The large sample can have a major impact on the heterogeneity, noise accumulation, spurious correlation, and incidental endogeneity.

The heterogeneity stems from the fact that big data are often created via aggregating many data sources corresponding to different subpopulations, and it is more likely that each subpopulation might exhibit some unique features not shared by other subpopulations. Using traditional statistical techniques may not be enough to analyze the data. Again, data characteristics will also have a major influence in mixture modeling compared with traditional data sets.

Noise accumulation occurs since big data analysis requires simultaneously estimating or testing many parameters. The estimation errors will compound as the data grows. Spurious correlation is evident in big data, where there are

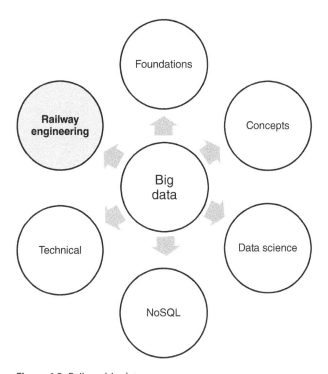

Figure 4.2 Railway big data

cases and situations in which many uncorrelated random variables may have high sample correlations in high dimensions. This can lead to incorrect statistical inferences. Incidental endogeneity refers to the genuine existence of a correlation between variables unintentionally due to high dimensional data. This endogeneity can cause inconsistency in model selection.

Figure 4.2 shows a schematic example of big data applications in railway engineering.

Big data is about extremely large volumes of data originating from various sources, such as databases, audio and video files, millions of sensors, and other systems. The sources of data in some cases provide outputs that are structured, but most are unstructured, semi-structured, or poly-structured. Furthermore, these data are streaming in some cases at a high velocity, and the data exposes at a higher speed or some speed as it is generated. The main key to the application of the big data paradigm relies heavily on the selection of appropriate data science techniques.

Hu et al. (2014) presented an overview of big data analytics. The authors summarized three definitions of big data:

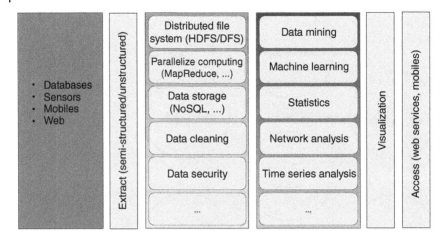

Figure 4.3 Big data environment

1) *Attribute definition* – This defines big data technologies as "a new generation of technologies and architectures, designed to economically extract value from very large volumes of a wide variety of data by enabling high velocity capture, discovery, and/or analysis" (Cooper and Mell, 2012).
2) *The second definition is more subjective* – Big data consists of "data sets whose size is beyond the ability of typical database software tools to capture, store, manage, and analyze." This is based on the McKinsey report (Manyika et al., 2011).
3) *The final definition that is often referred to as the architectural definition* – "Big data is where the data volume, acquisition velocity, or data representation limits the ability to perform effective analysis using traditional relational approaches or requires the use of significant horizontal scaling for different processing" (Cooper and Mell, 2012).

Figure 4.3 shows the big data environment.

Relational databases may be difficult to use when processing big data. Table 4.1 shows some applications of appropriate system platforms for large data sets.

4.2 Query

Requiring information from big data involves various operations, such as "joins" or "data filtering." There are various query systems designed for such purposes.

Table 4.2 shows the comparison between big data and traditional data.

Table 4.1 System for large data applications.

System	Data model
BigTable	Column families
Dynamo	Key-value storage
Hbase	Column families
Cassandra	Column families
Hypertable	Multidimensional table
MongoDB	Document-oriented storage
CouchDB	Document-oriented storage

Table 4.2 Comparison between big data and traditional data.

	Traditional data	Big data
Volume	Gigabytes (GB)	Constantly updated (terabytes (TB) or petabytes (PB) currently)
Generated rate	Per hour, day, …	More rapid
Structure	Structured	Semi-structured or unstructured
Data source	Centralized	Fully distributed
Data integration	Easy	Difficult
Data store	RDBMS	HDFS, NoSQL
Access	Interactive	Batch or near real time

Wu presented a survey of large-scale data management systems for big data applications. The outcome divided data management systems into two broad groups:

1) Relational
2) Non-relational

Figure 4.4 depicts the landscape. The non-relational zone involves the ongoing updates and improvements of infrastructures currently in the big data research. The authors also presented the taxonomy of data model (Figure 4.5).

Kune et al. (2016) presented the major differences between traditional databases and big databases. Figure 4.6 is a modified figure, and Table 4.3 illustrates the differences.

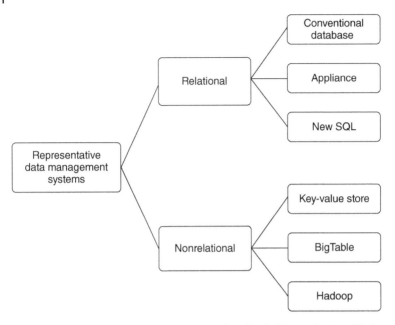

Figure 4.4 Landscape (Wu et al. (2015). Reproduced with the permission of Springer)

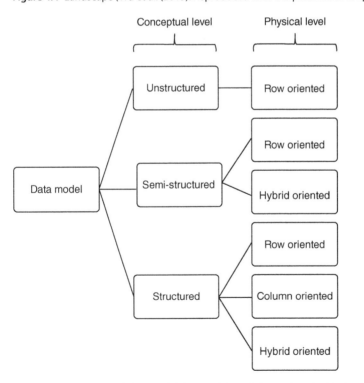

Figure 4.5 Taxonomy of data model (Wu et al. (2015). Reproduced with the permission of Springer)

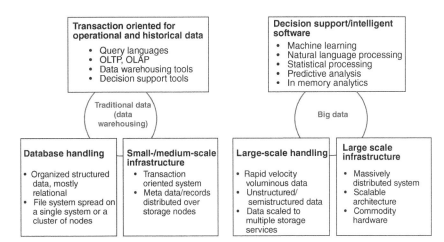

Figure 4.6 Big data versus traditional data (Kune et al. (2016). Reproduced with the permission of John Wiley and Sons)

Figure 4.7 shows the big data taxonomy.

The big data analytics can be grouped into two alternative paradigms that are present in railway track engineering:

1) *Stream processing* – The potential value of data depends on data freshness. The major characteristic is that data arrives in a stream; it is continuous, and only a limited portion can be stored.
2) *Batch processing* – In this application, the data are stored and analyzed later. In some cases, the data are analyzed in subsets.

Table 4.4 compares streaming processing and batch processing.

The development of advanced sensors and information technologies in critical infrastructure monitoring and control has provided a platform for the expansion and growth of data. This has created a new paradigm in the processing, storing, streaming, and visualization of this data and information. The changes in technology include the possibilities of installing sensors and smart chips in critical infrastructure to measure system performance, physical characteristics, and other indicators of imminent failures. Furthermore, many of the critical infrastructure components contain communication capabilities (e.g., IP addresses and direct Internet or wireless connections to the Internet) that will allow data to be uploaded automatically or on demand (Meeker and Hong, 2013).

As an emerging terminology, big data is an important paradigm that should be considered in a resilience engineering application. Big data can be formally

Table 4.3 Traditional data warehousing versus big data issues.

Property	Traditional data warehousing	Big data-specific issues
Data volume	Data are segregated into operational and historical data. Apply extraction, transformation, and load mechanisms for processing. As the data volumes are increased, the historical data are filtered from the warehouse system for faster database queries	High volume of data from several sources like the web, sensor networks, social networks, and scientific experiments. Capable of handling operational and historical data together, which could be replicated on multiple storage devices for high availability and throughput
Speed	Transaction-oriented and the data in turn generated from the transactions are low	High data growth due to several sources like web and scientific sensors streaming experiments
Data formats	Semi-structured/structured data like XML and relational	Multi-structured data handling, such as relational, and unstructured/semi-structured, such as text, XML, video streaming, and so on
Applicable platforms	Online transnational processing, relational database management system	Big data analytics, text mining, video analytics, web log mining, scientific data exploration, intrinsic information extractions, graph analytics, social networking, in-memory analytics, and statistical and predictive analytics
Programming methodologies/languages	Query language like SQL	Data-intensive computing languages for batch processing and stream computing, like MapReduce and NoSQL programming
Data backup/archival	Files/relational data need to have data backup procedures or mechanisms. Traditional data works on regular, incremental, and full backup mechanisms that are already established	Due to large and high speeds of the data growth rates, the conventional methods are not adequate; hence, techniques such as differential backup mechanisms need to be explored
Disaster recovery (DR)	Data are replicated at several places to address the disaster	DR techniques could be separated from mission critical and noncritical data
Relationship with clouds	Relational databases/data warehousing tools as services over cloud infrastructures	On-demand big data infrastructure setup, analytic services by several cloud, and big data providers
Data deduplication	Applicable to transactional record deduplication while merging database records	File and block level deduplication mechanisms need to be explored for continuous growing and stream-oriented data
System users	Administrators, developers, and end users	Data scientists and analytics end users

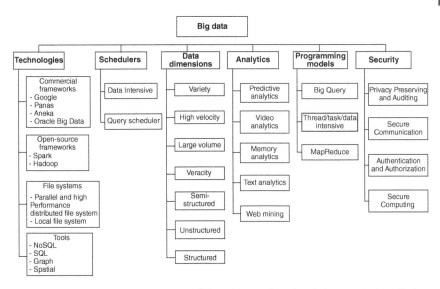

Figure 4.7 Big data taxonomy (Kune et al. (2016). Reproduced with the permission of John Wiley and Sons)

Table 4.4 Comparison between stream processing and batch processing.

	Stream processing	Batch processing
Input	Stream of new data or updates	Data chunks
Data size	Infinite or unknown in advance	Known and finite
Storage	Not stored or only a nontrivial portion is stored	Stored
Hardware	Typically a single limited amount of memory	Multiple CPUs, memories
Processing	A single or few pass(es) over data	Processed in multiple rounds
Time	A few seconds or even milliseconds	Much longer
Applications	Web mining, sensor networks, traffic monitoring	Widely adopted in almost every domain

defined as a collection of huge, diverse data sets that make it practically impossible to analyze and draw inferences using traditional data processing platforms. With diversified data provisions in critical infrastructures, like sensors, scientific experiments, and high throughput instruments, the amount of data generated has grown exponentially.

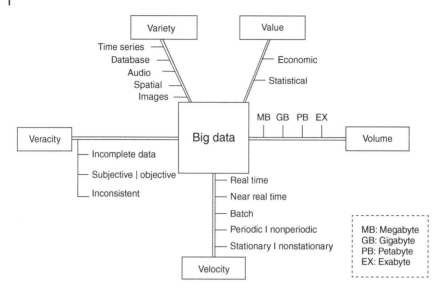

Figure 4.8 Five Vs of big data

Big data consists of multidimensional, multimodal data sets and can be characterized by five Vs. Figure 4.8 shows the five Vs.

- *Volume* – This refers to the huge amount of data that needs to be processed, stored, and analyzed. This requires developing algorithms that are scalable.
- *Velocity* – This is the indication of how quickly data is to be analyzed. Some challenges include processing the data in near real time or virtually real time.
- *Variety* – This is related to the different types of structured or unstructured data. The integration of different types of data is paramount.
- *Veracity* – This is the indication of the integrity of the data. Therefore, a robust and predictive algorithm capable of handling noise and incomplete data and information is needed for decision-making.
- *Value* – This is an important feature in big data. It mainly refers to the worth of the information being managed. Value also includes the discovery of knowledge and high return of investment.

The four Vs are concerned about data collection, preprocessing, transmission, and storage. The V that represents value focuses on extracting value from the data using statistical and machine learning algorithms (Ang and Seng, 2016). The future perspective and challenges include the following:

- Other than the five Vs, one has to consider *E* (energy efficiency) to be fulfilled.
- Most of the big data systems are focused on the "volume" characteristics. More emphasis is needed on some of the other Vs, such as variety.

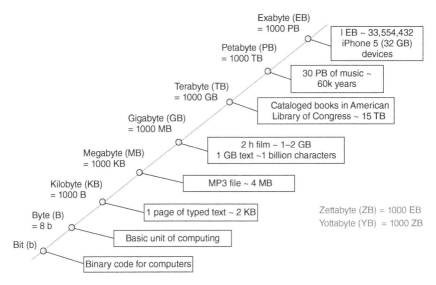

Figure 4.9 Data size (Adarkwa, 2015). Reproduced with the permission of University of Delaware

- Traditional, mathematical, and simulation techniques fail to determine current parameters and other factors.
- Machine learning techniques and the emergence of deep learning techniques appear to have more promise in the big data systems.

Some big data challenges include:

- Quality versus quantity and questions like the following:(a) How do we decide which data is irrelevant versus selecting the most relevant data? (b) How reliable and accurate is the data? (c) How much data is enough to make accurate predictions or to determine the correct probability distribution?
- Speed versus scale
- Structured versus unstructured data
- Compliance and security
- Distributed data and distributed processing (Kaisler et al., 2013). Figure 4.9 provides an idea of data size.

4.3 Taxonomy of Big Data Analytics in Railway Track Engineering

In railway track engineering, appropriate big data technology is needed to obtain the best solutions (Cloud Security Alliance, 2014). Issues include (a) event characteristics, which include input/output data required for specific infrastructures or a combination of different infrastructures, and (b) event response complexity, which involves processing complexity, data domain complexity, and size/storage of the data. Also, it is very important to understand the latency requirement. In resilience engineering, it is more appropriate to address "low latency" applications (a response time on the order of a few tens of milliseconds). The "medium to high latency" are those that need a response on the order of a few seconds to minutes. In civil infrastructure systems, one may also consider "higher" latency.

Table 4.5 is an example of big data taxonomy in railway engineering.

The next generation of railway track engineering data has all the characteristics of big data (Table 4.6); for example, sensors embedded to monitor track infrastructure will provide streams of data and unstructured data (Table 4.7).

The big data paradigm will change greatly how railway track engineering data is analyzed. Today, railway track engineering data in the era of big data can be illustrated using Table 4.8.

Jin et al. (2015) highlighted some of the challenges in harnessing the potential of big data, including:

- *Data complexity* – The coherent characteristics and their perception, representation, understanding, and computation can be a little problematic. Proper formulation is therefore essential in addressing some of the complexities.
- *Computational complexity* – The multiple sources of data, different characteristics, and, in some cases, streaming make the use of conventional computational techniques obsolete. One of the main keys to address this challenge is to focus on more data-centric computing and to study nondeterministic algorithm methods.
- *System complexity* – Energy efficiency is most critical in addressing the system complexity issues.

Tsai et al. (2015) noted that most data analysis methods have limitations for big data:

a) *Unscalability and centralization* – In most cases, the regular data analysis techniques are for large-scale and complex data sets. They usually fail to scale up.
b) *Non-dynamic* – The traditional data analysis methods in some occasions cannot be dynamically adjusted for different situations.
c) *Uniform data structure* – Traditional data analysis has been more appropriate in these situations.

4.4 Data Engineering

Big data analysis requires engineers to manage an immense amount of data in a relatively short time. This has led to the development of tools for large data

Table 4.5 Taxonomy of big data methods in railway track engineering.

Analysis domain	Sources	Characteristics	Approaches	Comments
Structured data	Field data collection Sensors Scientific experiment data	Structured records Real time	Data mining Statistical analysis	All infrastructure systems need field data
Unstructured data	Extreme events Sensors	Unstructured records Mixture of variables	Anomaly detection	Infrastructure inspection reports Specification updates
Text analytics	Logs	Unstructured	Document presentation	Early detection
	Email	Rich textual	NLP	
	Corporate documents	Context	Information extraction	
	Government rules and regulations	Semantic	Topic model	
	Text content of webpages Citizen feedback and comments	Language dependent	Summarization Categorization Clustering Question answering Option mining	
Multimedia analytics	Corporation-produced multimedia	Image, audio, video	Summarization	Early detection
	User-generated multimedia		Annotation	
	Surveillance	Massive Redundancy Semantic gap	Indexing and retrieval Recommendation Event detection	
Mobile analytics	Mobile apps Sensors		Monitoring Location-based mining	
	RFID			

Table 4.6 Key definitions of railway track engineering.

Terms	Definitions and examples
Entity	Key units of analysis or objects of research interest, including: • Devices: RFID tags, GPS locators
Data	Attributes, features, actions, and events identified to describe the entity in various kinds of settings: • Contextual • Spatial • Temporal
Data agency	Institutions that collect and store relevant data in different settings

Table 4.7 Data definition.

Type	Definition
Micro-data	The least aggregated level of data in very large data sets
Meso-data	The mid-level of data aggregation in very large data sets, resulting from the collection of data that are a level up from micro-data in terms of the kinds of information that is captured
Macro-data	The most aggregated level of data in large data sets that describes regional or geographic areas

Table 4.8 Railway track engineering and big data.

Current railway track engineering data	Railway track engineering data (big data era)
Rear-view mirror hindsight	"Forward-looking" recommendations
Relatively little data available	Exploit all data from independent sources
Batch, incomplete, disjointed	Real time
Monitoring	Optimization

analysis. Apache Hadoop is an open-source software framework that supports data-intensive, naively distributed, and natively parallel applications. Hadoop implements a computational algorithm called MapReduce. Hadoop also provides a distributed file system – Hadoop File Systems (HDFS) – that stores data on the computer nodes, facilitates rapid data transfer rates among nodes, and allows the systems to continue operating uninterrupted in the case of node failure (Schmarzo, 2013). Apache Hadoop is the most popular implementation of MapReduce (Fernández et al., 2014).

MapReduce is a programming architecture proposed by Google for a distributed processing of large data sets on massively parallel systems. The MapReduce contains two main phases: a map function and a reduce function. The map and reduce functions work as follows (Fernández et al., 2014):

- *Map function* – The master node performs a segmentation of the input data set into independent blocks and splits them into worker nodes. The worker nodes then process the smaller problems and pass the answer back to the master node.
- *Reduce function* – The master node collects the answers to all the subproblems and combines them in some way to form the final output.

Figure 4.10 shows the MapReduce architecture. Therefore, a typical problem solved by MapReduce proceeds as follows (Antoniu and Fedak, 2010):

1) Read a lot of data.
2) Map-extract something you are interested in from each record.
3) Shuffle and sort.
4) Reduce-aggregate, summarize, filter, or transform.
5) Note the output.

The terms k and v refer to the key and value pairs, respectively. MapReduce has the following advantages (Lee et al., 2011): (a) simple and easy to use; (b) flexible and does not have any dependency on data model and schema; (c) independent of the storage; (d) fault tolerance; and (e) high scalability. There are some pitfalls: (a) no high-level language; (b) no schema and index; (c) a single fixed data flow, which makes complex algorithms hard to implement and analyze; (d) low efficiency; and (f) finally, it is still relatively new, so third party tools are relatively few. Although MapReduce has been used extensively,

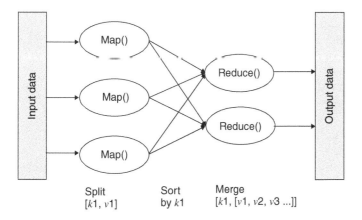

Figure 4.10 MapReduce architecture (Attoh-Okine, 2016). Reproduced with the permission of Cambridge University Press

iterative algorithms are not effective, and there are few alternative tools that have been developed.

Figure 4.11 shows the pseudocode of *k*-means clustering algorithms, and Figure 4.12 shows the corresponding MapReduce algorithm.

Spark is an alternative to Hadoop, which is designed to overcome the input/output limitations. Spark has the ability to perform in-memory computation. Spark has several interfaces including (Singh and Reddy, 2015):

- Scala (Java functional programming)
- Java
- Python

Big data addresses theoretical problems such as:

a) How do microscale factors influence macroscale phenomenon?

Factors of big data include:

a) Technical issues
b) Conceptual issues
c) Privacy issues

Input: Data points D, Number of clusters k
Step 1: Initialize k centroids randomly
Step 2: Associate each data point in D with the nearest centroid. This will divide the data points into k clusters
Step 3: Recalculate the position of centroids Repeat steps 2 and 3 until there are no more changes in the membership of the data points
Output: Data points with cluster memberships

Figure 4.11 Pseudocode of the *k*-means algorithm

K-means – Map
Input: Data points D, number of clusters k and centroids
Step 1: For each data point $d \in D$ do
Step 2: Assign d to the closest centroid
Output: Centroids with associated data points

K-means – Reduce
Input: Centroids with associated data points
Step 1: Compute the new centroids by calculating the average of data points in cluster
Step 2: Write the global centroids to the disk
Output: New centroids

Figure 4.12 Pseudocode MapReduce-based *k*-means algorithm

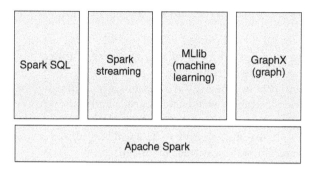

Figure 4.13 Apache Spark

The core data units Spark are called Resilient Distributed Datasets (RDDs) (Reyes-Ortiz et al., 2015). They are a distributed, immutable, and fault-tolerant memory abstraction that collects a set of elements on which a set of operators can be applied to either produce other RDDs (transformation) or return values (actions). The RDDs can reside in memory, in desk, or in combination. The RDD follows a lazy evaluation (LE) strategy, where minimal computations are performed, and unnecessary memory usage. The different components include (Figure 4.13):

- *Spark SQL* – Provides support to structure and semi-structured data
- *Spark streaming* – Used to perform streaming
- *Machine learning library (MLlib)* – Distributed machine learning framework
- *GraphX* – Is a distributed graph-processing framework

The technical issues for big data applications in railway track engineering center on data collection, storage, retrieval, management, data privacy, and analysis. Most railway operators and agencies collect data from multiple sources – geometry cars, rail defect cars, and subsurface investigation using ground-penetrating radar. Most of these data are collected in different formats.

Currently, there are improved technologies that allow track operators to collect data for large numbers of track miles in relatively short periods, significantly enlarging the volume, velocity, and variety of data available for analysis and hence maintenance decisions.

The size of storage and retrieval of track data is going to be a major problem for railway agencies – especially in the case of retrieval. The railway agencies are collecting and storing large data sets. Therefore, the use of repositories still needs to be addressed. In terms of analysis, for example, images are usually transformed in quantitative values both for further input and interpretation. The correct interpretation will almost always depend on signal processing algorithms or the tool used. The conceptual and theoretical issues included are that big data may have a major influence on the sample data selected for analysis and inference:

- How can predictive analysis (including standards and specifications) not developed based on large data be compared?
- Privacy issues are very important, especially in cases that involve derailments and accidents. The major question railway agencies need to address is to how agencies will protect their huge amounts of data. But it is also very important to remind railway operators that non-sharing of data will limit the advancement of the big data paradigm in railway track engineering.

4.5   Remarks

The future of railway track engineering data modeling and analyses is definitely going to change in the new era of big data. There is a need to introduce data science and big data paradigm to assist railway track engineering in analyzing huge data sets; hence, better safety and maintenance decisions can be reached.

References

O. A. Adarkwa. *Tensor factorization in civil infrastructure systems*. PhD thesis, University of Delaware, 2015. http://gradworks.umi.com/37/30/3730220.html.

L.-M. Ang and K. P. Seng. Big sensor data applications in urban environments. *Big Data Research*, 4:1–12, 2016. doi: 10.1016/j.bdr.2015.12.003.

G. Antoniu and G. Fedak. Scalable distributed processing using the map-reduce paradigm, 2010. https://www.grid5000.fr/mediawiki/images/101005-Hemera-Challenge-MapReduce.pdf.

N. Attoh-Okine. *Resilience Engineering: Models and Analysis*. Cambridge University Press, 2016. http://www.cambridge.org/us/academic/subjects/engineering/engineering-mathematics-and-programming/resilience-engineering-models-and-analysis?format=HB&isbn=9780521193498; https://books.google.com/books?id=O_-lCwAAQBAJ&pg=PR1&lpg=PR1&dq=Resilience+engineering:+models+and+analysis+attoh-okine&source=bl&ots=pd9zBxkGAv&sig=55K7-scwsEVae1oN5ZvBwTH4vNk&hl=en&sa=X&ved=0ahUKEwiA87rs0djRAhVG4iYKHWZrDh0Q6AEIYDAH#v=onepage&q=Resilience%20engineering%3A%20models%20and%20analysis%20attoh-okine&f=false.

Cloud Security Alliance. Big data taxonomy. Technical report, 2014. https://downloads.cloudsecurityalliance.org/initiatives/bdwg/Big_Data_Taxonomy.pdf.

M. Cooper and P. Mell. Tackling big data, 2012. http://csrc.nist.gov/groups/SMA/forum/documents/june2012presentations/fcsm_june2012_cooper_mell.pdf.

J. Fan, F. Han, and H. Liu. Challenges of big data analysis. *National Science Review*, **1**(2):293–314, 2014. doi: 10.1093/nsr/nwt032.

A. Fernández, S. del Río, V. López, A. Bawakid, M. J. del Jesus, J. M. Benítez, and F. Herrera. Big data with cloud computing: an insight on the computing environment, MapReduce, and programming frameworks. *Wiley Interdisciplinary Reviews: Data Mining and Knowledge Discovery*, **4**(5):380–409, 2014. doi: 10.1002/widm.1134.

H. Hu, Y. Wen, T.-S. Chua, and X. Li. Toward scalable systems for big data analytics: a technology tutorial. *IEEE Access*, **2**:652–687, 2014. doi: 10.1109/ACCESS.2014.2332453.

D. Hunt, J. Kuehn, and O. Wyman. Big data and railroad analytics, 2012. http://blogs1.oliverwyman.com/rail/wp-content/uploads/sites/4/2012/02/Big-Data_RAS-Newsletter-2011-12.pdf.

H. V. Jagadish. Big data and science: myths and reality. *Big Data Research*, **2**(2):49–52, 2015. doi: 10.1016/j.bdr.2015.01.005.

X. Jin, B. W. Wah, X. Cheng, and Y. Wang. Significance and challenges of big data research. *Big Data Research*, **2**(2):59–64, 2015. doi: 10.1016/j.bdr.2015.01.006.

S. Kaisler, F. Armour, and J. A. Espinosa. Big data: issues and challenges moving forward. In *46th Hawaii International Conference on System Sciences*, 2013. http://www.computer.org/csdl/proceedings/hicss/2013/4892/00/4892a995.pdf.

R. Kune, P. K. Konugurthi, A. Agarwal, R. R. Chillarige, and R. Buyya. The anatomy of big data computing. *Software: Practice and Experience*, **46**(1):79–105, 2016. http://dl.acm.org/citation.cfm?id=2904648.2904654.

K.-H. Lee, Y.-J. Lee, H. Choi, Y. D. Chung, and B. Moon. Parallel data processing with MapReduce: a survey. *SIGMOD Record*, **40**(4), 2011. http://www.cs.arizona.edu/ bkmoon/papers/sigmodrec11.pdf.

J. Manyika, M. Chui, B. Brown, J. Bughin, R. Dobbs, C. Roxburgh, and A. H. Byers. Big data: the next frontier for innovation, competition, and productivity. Technical Report, 2011. http://scholar.google.com/scholar.bib?q=info:kkCtazs1Q6wJ:scholar.google.com/&output=citation&hl=en&as_sdt=0,47&ct=citation&cd=0.

W. Q. Meeker and Y. Hong. Reliability meets big data: opportunities and challenges. *Quality Engineering*, **26**(1):102–116, 2013. doi: 10.1080/08982112.2014.846119.

J. L. Reyes-Ortiz, L. Oneto, and D. Anguita. Big data analytics in the cloud: spark on Hadoop vs MPI/OpenMP on Beowulf. *Procedia Computer Science*, **53**:121–130, 2015. doi: 10.1016/j.procs.2015.07.286.

B. Schmarzo. *Big Data: Understanding How Data Powers Big Business*. Wiley, 2013. ISBN: 1118740009. https://books.google.com/books?id=Tez9AAAAQBAJ&pgis=1.

D. Singh and C. K. Reddy. A survey on platforms for big data analytics. *Journal of Big Data*, **2**(1):1–20, 2015. doi: 10.1186/s40537-014-0008-6.

C.-W. Tsai, C.-F. Lai, H.-C. Chao, and A. V. Vasilakos. Big data analytics: a survey. *Journal of Big Data*, **2**(1):21, 2015. doi: 10.1186/s40537-015-0030-3.

L. Wu, L. Yuan, and J. You. Survey of large-scale data management systems for big data applications. *Journal of Computer Science and Technology*, **30**(1):163–183, 2015. doi: 10.1007/s11390-015-1511-8.

5

Hilbert–Huang Transform, Profile, Signal, and Image Analysis

5.1 Hilbert–Huang Transform

The Hilbert–Huang transform (HHT) is a very powerful tool for data analysis due to its ability to process nonstationary data. The HHT consists of two key components: empirical mode decomposition (EMD) and Hilbert spectral analysis. A summary of the process is illustrated in Figure 5.1.

The first component of the HHT is the EMD. The essence of EMD is to decompose different trends or fluctuations contained in the complicated signal gradually through their characteristic scales to obtain a series of data sequences with different intrinsic timescales, that is, the intrinsic mode functions (IMFs). This decomposition helps create a better understanding of the internal structure of the signal and the components involved (Klionski et al., 2008). The EMD algorithm is highly efficient and adaptive as it preserves the nonstationary and nonlinear characteristics in the IMFs (basis functions) through its local wave analysis property (Zhou et al., 2009). These IMFs mainly possess two important properties ((Huang et al., 1998); (Klionski et al., 2008)):

- The number of extrema and the number of zero crossings must either equal or differ by no more than one in the whole data set:

$$N_{max} + N_{min} = N_{zero} \pm 1, \tag{5.1}$$

where N_{max} = total number of maxima, N_{min} = total number of minima, and N_{zero} = total number of zero crossings.

- At any point, the mean $m(t)$ of the envelope defined by the local maxima and the envelope defined by the local minima is zero:

$$\frac{E_{max}(t) + E_{min}(t)}{2} \approx 0, \tag{5.2}$$

where $E_{max}(t)$ = envelope of the local maxima and $E_{min}(t)$ = envelope of the local minima by spline interpolation.

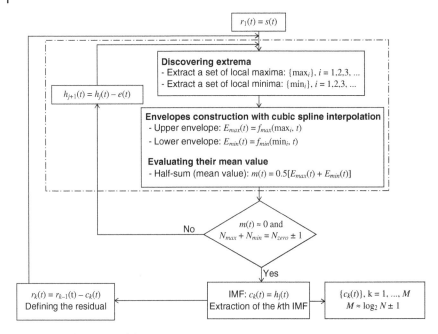

Figure 5.1 Illustration of the HHT

5.1.1 Traditional Empirical Mode Decomposition

The empirical modes for the traditional EMD are extracted from a complicated data set by the following process. First, identify all the local extrema (maxima $-\{max\}$ and minima $-\{min\}$) of the input signal $X(t)$ where t belongs to $[1, \ldots, N]$. Then connect all the local maxima by a cubic spline to produce the upper envelope $E_{max}(t)$ as

$$E_{max}(t) = f_{max}(max_i, t). \tag{5.3}$$

The procedure is repeated for the local minima to produce the lower envelope $E_{min}(t)$ as

$$E_{min}(t) = f_{min}(min_i, t). \tag{5.4}$$

The upper and lower envelopes should encompass all the data between them. This implies

$$\min(t) \le X(t) \le \max(t). \tag{5.5}$$

The mean of the upper and lower envelopes is designated as

$$m_1(t) = \frac{E_{min}(t) + E_{max}(t)}{2}, \tag{5.6}$$

and the difference between the data $X(t)$ and $m_1(t)$ is the first component h_1:

$$X(t) - m_1(t) = h_1(t). \tag{5.7}$$

Ideally, $h_1(t)$ should be an IMF. Errors might be introduced due to the spline-fitting process. This calls for repetition of the previous process, normally referred to as a "sifting process." Figure 5.2 shows the sifting process. In the subsequent sifting processes, $h_1(t)$ is treated as the data and a new mean is computed:

$$h_{11}(t) = h_1(t) - m_{11}(t) \tag{5.8}$$

Here, m_{11} = mean of the upper and lower envelopes for h_1. Repeat the sifting process up to k times; h_{1k} becomes an IMF:

$$h_{1k}(t) = h_{1(k-1)}(t) - m_{1k}(t). \tag{5.9}$$

So, h_{1k} becomes the first IMF. To reduce the complexity of the following equations, replace h_{1k} with $C_j, j = 1, \dots, n$. C_1 = is the highest frequency component of the signal. C_1 is removed from the original data to obtain a residue:

$$r_1(t) = X(t) - C_1(t). \tag{5.10}$$

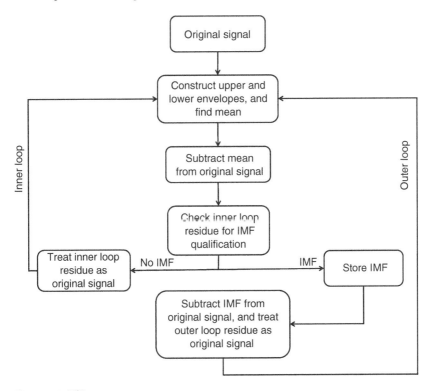

Figure 5.2 Sifting process

The residue, r_1, is treated as new data, and the sifting process is repeated as described above. The sifting procedure is repeated for all subsequent r_i functions as

$$r_{j-1} - C_j = r_j, \quad j = 2, 3, \dots, n. \tag{5.11}$$

The process is terminated if the residue becomes monotonic (only one extrema). The original data are thus the sum of the IMF components plus the final residue:

$$X(t) = \sum_{j=1}^{n} C_j(t) + r_n(t) \tag{5.12}$$

C_jth IMFs and n = number of sifted IMF in the above equation. r_n represents either the mean trend or a constant. Oversifting can smooth the amplitudes of IMFs, rendering them physically less meaningful. Oversifting is avoided by limiting the size of the sum of the difference (SD), computed from two consecutive sifting results as

$$\mathrm{SD} = \frac{\sum_{t=0}^{T} \left| h_{k-1}(t) - h_k(t) \right|^2}{\sum_{t=0}^{T} h_{k-1}^2(t)}. \tag{5.13}$$

Figure 5.3 is a sample of synthetic signal data for which the application of the HHT is applicable and capable of identifying the major trends and noise.

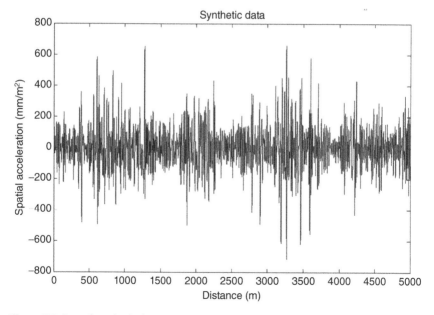

Figure 5.3 Part of synthetic data

Figure 5.4 shows the IMF components resulting from EMD of the synthetic data.

Figure 5.5 is the plot of the instantaneous wave number against distance for the highest wave number component IMFs.

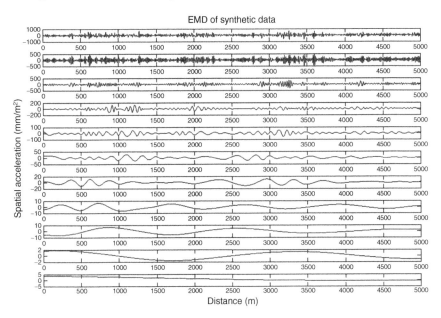

Figure 5.4 Signal and IMF components

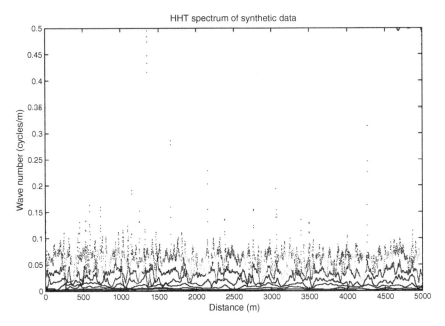

Figure 5.5 Plot of instantaneous wave number against distance for highest wave number component IMFs

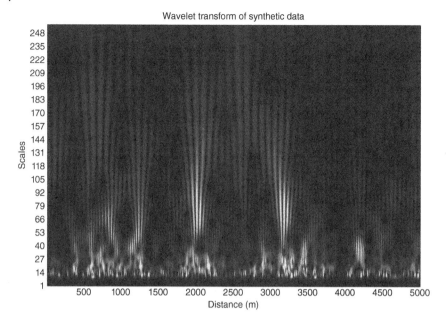

Figure 5.6 Wavelet transform of synthetic data

Figure 5.6 shows the wavelet transform of the same signal.

5.1.1.1 Side Effect (Boundary Effect)

The boundary effect points affect the shape of the spline function. This may lead to different IMFs. The following methods are used to address the boundary effect:

- *Wave extending method*. Using standard EMD procedure with two consecutive local maxima points to predict the boundary value.
- *Mirror extending method*. This uses mirror synchronization to create a new sequence.
- *Data extending method*. Extract data $X(i)$ with $i = 1, 2, \dots, N$. Start the following procedure (Ding and Lin, 2010).

  Create an even extension series $X_e(i)$ to generate a sequence with its frequency as $1/2N$. The sequence can be illustrated below:

$$X_e(i) = \begin{cases} X(i) & 1 \le i \le N \\ X(2N - i + 1) & N + 1 \le i \le 2N \, . \\ X(1) & i = 2N + 1 \end{cases} \tag{5.14}$$

Create maximum even series $X_e(i)$; if the right boundary point is not a local maximum point, set the leftmost maximum value as the boundary value. Apply cubic interpolation to all the maximum points and calculate the upper envelope. Do the same process for calculating the lower envelope. Create an odd extension series $X_0(i)$, and generate the $2N$ sequence as well:

$$X_0(i) = \begin{cases} X(i) & 1 \leq i \leq N \\ -X(2N - i + 1) & N + 1 \leq i \leq 2N \\ X(i) & i = 2N + 1 \end{cases} \quad (5.15)$$

Repeat the previous process for the upper envelope and the lower envelope of $X_0(i)$. Use the mean of the four envelopes as the mean spline value:

- *Data reconstruction method (Ding and Lin, 2010).*
  Find all local maximum and local minimum. List the maximum points and the minimum points to matrix form, $\max = [x(1) \max x(N)]$ and $\min = [x(1) \min x(N)]$.

- *Similarity searching method.* This method uses a moving time window to separate the original signal to several parts X_i and can be described below:

$$X_i = [x(i), x(i + 1), \dots, x(i + w - 1)]^T, \quad (5.16)$$

where w is the length of the moving time window.
Then use the neighborhood searching method to find the most similar vector with the boundary points:

$$X_{nearest} = \underset{i}{\arg\min} \|X_i - X_{endpoint}\|. \quad (5.17)$$

For the first boundary point, the subseries contain one maximum point and one minimum point. We adapt the two points in the front and the end of the series for extending. Then, follow the standard procedure.

5.1.1.2 Example
Figures 5.7–5.12 show the application of the HHT to cross-level, surface, and alignment. The analysis depicts the trend and behavior of different variables at different distances.

5.1.1.3 Stopping Criterion
The algorithm was proposed by Junsheng et al. (2006). Assuming $x(t)$ contains several independent orthogonal components $x_i(t)$,

$$x(t) = x_1(t) + x_2(t), \dots, x_n(t) = \sum_{i=1}^{n} x_i(t). \quad (5.18)$$

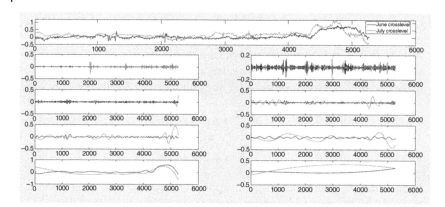

Figure 5.7 Analysis of cross-level

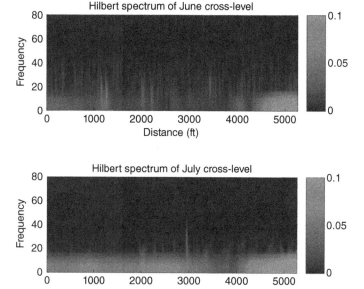

Figure 5.8 Comparative analysis of cross-level at different months (June and July)

The total energy of the original signal $x(t)$ can be represented as the following equation:

$$E_x = \int_{-\infty}^{\infty} x^2(t)dt = \int_{-\infty}^{\infty} \left[\sum_{i=1}^{n} x_i(t)^2 dt \right].$$ (5.19)

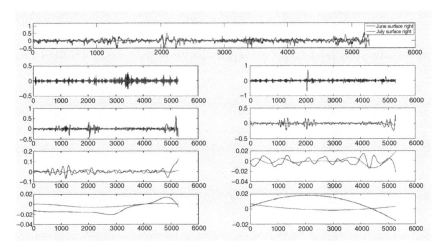

Figure 5.9 Analysis of surface (right)

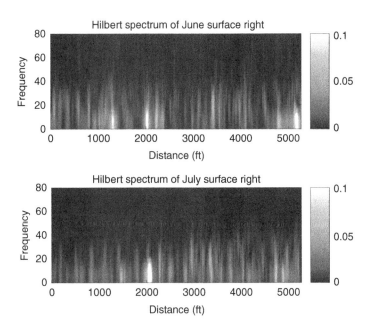

Figure 5.10 Comparative analysis of surface (right)

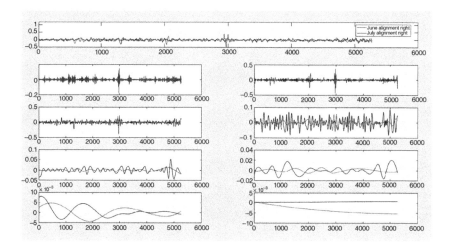

Figure 5.11 Analysis of alignment (right)

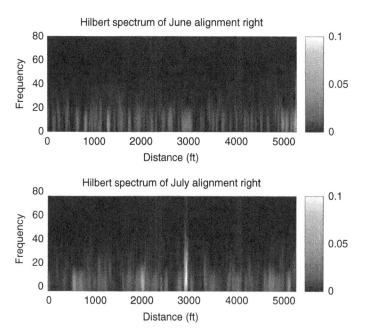

Figure 5.12 Comparative analysis of alignment (right)

Since we assume $x(t)$ can be separated into several orthogonal components, the energy can be represented as the following equation:

$$E_{tot} = \int_{-\infty}^{\infty} c_1^2(t)dt + \int_{-\infty}^{\infty} \left[x(t) - c_1(t)\right]^2 dt$$

$$= 2E_{c_1} + E_x - 2\int_{-\infty}^{\infty} x(t)c_1(t)dt \tag{5.20}$$

Use the energy difference between E_x and E_{tot} to decide the criterion.

5.1.2 Ensemble Empirical Mode Decomposition (EEMD)

A major drawback of the traditional EMD described above is a result called mode mixing. Mode mixing results when the decomposition method is unable to collate signals with similar frequencies into each IMF. As a result, different modes of oscillations reside in each IMF, making IMFs lose physical meaning or falsely represent the physical processes in a mode. This drawback resulted in the introduction of the ensemble empirical mode decomposition (EEMD), a noise-assisted data analysis method. Below is a summary of the EEMD process ((Wu and Huang, 2009); (Lei et al., 2009)):

1) Initialize the number of ensemble M, the amplitude of the added white noise, and $m = 1$.
2) Perform the mth trial on the signal-added white noise:
   a) Add white noise series with the given amplitude to the investigated signal:

   $$x_m(t) = x(t) + \eta_m(t), \tag{5.21}$$

   where η_m indicates the mth added noise series and $x_m(t)$ represents the noise-added signal of the mth trial.
   b) Decompose the noise-added signal $x_m(t)$ into I IMFs $c_{i,m} (i = 1, 2, \dots, I)$ using the EMD method described in the previous section, where $c_{i,m} = i$th IMF of the mth trial and $I =$ number of IMFs.
   c) If $m < M$, then go to Step (a) with $m = m + 1$. Repeat Steps (a) and (b) multiple times but with a different white noise series each time.
3) Obtain an ensemble mean of the corresponding IMFs C_i of the decomposition, and

$$\overline{C_i(t)} = \frac{1}{M} \sum_{m=1}^{M} C_{i,m-1}(t) + r_m(t), \quad i = 1, 2, \dots, I; \quad m = 1, 2, \dots, M. \tag{5.22}$$

4) Report the mean $\overline{C_i(t)}$ of the IMFs as the final IMFs. The effect of the added white noise should decrease following the well-established statistical rule

$$\varepsilon_n = \frac{\varepsilon}{\sqrt{N}} \tag{5.23}$$

EEMD   Post-processig of EEMD

$$\mathrm{IMF}_1 = \frac{1}{m}\sum_{i=1}^{m}\mathrm{IMF}_{i1}$$ $$\mathrm{IMF}_1 \rightarrow \mathrm{pIMF}_1 + \mathrm{residual}_1$$

$$\mathrm{IMF}_2 = \frac{1}{m}\sum_{i=1}^{m}\mathrm{IMF}_{i2}$$ $$\mathrm{IMF}_2 + \mathrm{residual}_1 \rightarrow \mathrm{pIMF}_2 + \mathrm{residual}_2$$

$$\vdots$$ $$\vdots$$

$$\mathrm{IMF}_k + \mathrm{residual}_{k-1} \rightarrow \mathrm{pIMF}_k + \mathrm{residual}_k$$

$$\mathrm{IMF}_k = \frac{1}{m}\sum_{i=1}^{m}\mathrm{IMF}_{ik}$$ $$\Rightarrow X(t) = \sum_{j=1}^{k}\mathrm{pIMF}_k + \mathrm{trend}$$

Figure 5.13 Post-processing ensemble empirical mode decomposition. Courtesy: Ding and Lin, 2010

or

$$\ln \varepsilon_n + \frac{\varepsilon}{2}\ln N = 0, \tag{5.24}$$

where ε = amplitude of the added noise and ε_n = final standard deviation of error, which is defined as the difference between the input signal and the corresponding IMFs.

5.1.2.1 Post-Processing EEMD
The method will do EMD to all IMF. The procedure is shown in Figure 5.13.

5.1.3 Complex Empirical Mode Decomposition (CEMD)

The application of the traditional EMD and the EEMD to profile analysis is based on the assumption that a signal's oscillation is univariate. A new extension of the EMD to handle and analyze data that are intrinsically bivariate (complex valued) was introduced by Rilling and Flandrin (2008), Altaf et al. (2007), and Tanaka and Mandic (2007). The complex empirical mode decomposition (CEMD) is not a signal processing technique but rather a representation of exactly the original signal information, which may be more amenable to visual analysis than the original signal itself (Farnbach, 1975). The process is as follows:

1) Initialization: $k = 1$, project the complex-valued signal $x(t)$ onto a predetermined number of projection directions φ_n. The number of directions should satisfy

$$\varphi_n = 2n\pi/N, \quad n \in [1, N] \tag{5.25}$$

The projected signal is now represented as

$$p_{\varphi_n} = \mathrm{Re}(e^{-j\varphi_n}x_k(t)), \quad n \in [1, N] \tag{5.26}$$

2) Extract the maxima of $p_{\varphi_n}(t)$ and calculate the corresponding envelopes $e_{\varphi_n}(t)$ by spline interpolation in all projection directions.

3) Compute the mean of the envelopes in the different projection directions:

$$m_k(t) = \frac{1}{N} \sum_{n=1}^{N} e_{\varphi_n}(t) \tag{5.27}$$

4) Subtract the mean $m_k(t)$ from the original signal $x(t)$ to obtain $h_{h,i}(t)$:

$$h_{k,i}(t) = x_k(t) - m_k(t) \tag{5.28}$$

let $h_{k,i}(t)$ meet the stopping criteria; then it becomes the first complex IMF.

5) Record the obtained IMF. Let $x_k(t) = x(t)$, $k = k + 1$.

6) Repeat Steps 1–5 until K IMFs are calculated.

5.1.4 Spectral Analysis

The second and final step after the decomposition is based on the Hilbert transform. To transform from time-space data to time-frequency data, apply the Hilbert transform to each IMF, yielding instantaneous frequency and amplitude. For a given data, $X(t)$, the Hilbert transform, $Y(t)$, is defined as

$$Y(t) = \frac{1}{\pi} P \int_{-\infty}^{\infty} \frac{X(t)}{t - \tau} d\tau. \tag{5.29}$$

P represents the Cauchy principal value. With this definition, $X(t)$ and $Y(t)$ can be combined to form the analytical signal $Z(t)$, given by

$$Z(t) = X(t) + iY(t) = a(t)e^{i\theta(t)}, \tag{5.30}$$

$a(t)$ and $\theta(t)$ represent the amplitude and phase, respectively:

$$a(t) = \sqrt{X^2 + Y^2}. \tag{5.31}$$

$$\theta(t) = \tan^{-1}\frac{Y}{X}. \tag{5.32}$$

The instantaneous frequency can be defined as

$$\omega(t) = \frac{d\theta(t)}{dt} \tag{5.33}$$

After applying the Hilbert transform, each IMF is represented as

$$C_j = \text{Re}\left[a_j(t)e^{i\int w_j(t)dt}\right]. \tag{5.34}$$

This implies that the original equation will be expressed as

$$X(t) = \text{Re} \sum_{j=1}^{n} a_j(t)e^{i\int w_j(t)dt}. \tag{5.35}$$

Equation 5.29 is written in terms of the amplitude and instantaneous frequency associated with each component as functions of time. This differs from the time-independent amplitude and phase in the Fourier series representation of the following:

$$X(t) = \text{Re} \sum_{j=1}^{\infty} a_j e^{i w_j t}, \tag{5.36}$$

where a_i and $w_j =$ constants. This frequency time distribution of the amplitude is designated as the Hilbert amplitude spectrum $H(\omega, t)$, or simply the Hilbert spectrum. With the Hilbert spectrum defined, the marginal spectrum $h(\omega)$ is designated as

$$h(\omega) = \int_0^T H(\omega, t) dt. \tag{5.37}$$

The marginal spectrum offers a measure of total amplitude (or energy) contribution from each frequency value. It represents the accumulated amplitude over the entire data span in a probabilistic sense. Whereas the energy contribution at a particular frequency in the Fourier spectrum implies that a component of a sine or a cosine wave persists over the entire span of the signal, energy in the marginal spectrum at a certain frequency means only that, in the entire span of the signal, there is such a wave appearing locally ((Rudi, 2010)):

$$H(\omega, t) = H[\omega(t), t] := \begin{cases} a_1 \text{ on the curve } \{[\omega_1(t), t] : t \in \mathbb{R}\} \\ \vdots \\ a_n \text{ on the curve } \{[\omega_n(t), t] : t \in \mathbb{R}\} \\ 0 \text{ elsewhere} \end{cases}. \tag{5.38}$$

5.1.5 Bidimensional Empirical Mode Decomposition (BEMD)

The potential of the 1-D EMD generated research interest in 2-D applications for image processing. Existing traditional methods are still Fourier based, and processing is global rather than local so that essential information may be lost in the image during processing. To avoid loss of information, a 2-D version of the EMD has been recently developed. Algorithms have been developed in the literature to do 2-D sifting for bidimensional empirical mode decomposition (BEMD) ((Damerval et al., 2005); (Linderhed, 2005); (Nunes et al., 2005)), and they generally follow the process for the 1-D case, only modified to handle 2-D signals.

Linderhed (2005) first introduced the EMD in two dimensions, which is now popularly called the bidimensional empirical mode decomposition (BEMD).

BEMD was used for image compression, using only the extrema of the IMFs in the coding scheme. Nunes et al. (2005) developed a BEMD method for texture image analysis; BEMD was used for feature extraction and image filtering. The sifting process used is as follows:

- Identify extrema of the image, I, by morphological reconstruction based on geodesic operators.
- Generate the 2-D envelope by connecting the maxima points with the radial basis function (RBF).
- Determine the local mean, m_i, by averaging the two envelopes.
- Do $I - m_i = h_i$.
- Repeat the process.

For the envelope construction, the authors used the RBF of the form

$$s(x) = p_m(x) + \sum_{i=1}^{N} \lambda_i \Phi \left(\|x - x_i\| \right), \qquad (5.39)$$

where

- p_m is a low degree polynomial of the mth degree polynomial and d variables
- $\|\cdot\|$ denotes the Euclidean norm
- λ_i are RBF coefficients
- x_i are the RBF centers

The stopping criterion used is similar to that developed by Huang et al. (1998), using standard deviation. Linderhed (2005) also developed a sifting process for a 2-D time series. Although the stopping criterion for IMF extraction is relaxed, the stopping criterion for the whole EMD process is similar to that of Huang et al. (1998). The IMF stopping criterion is based on the condition that the extrema envelope is close enough to zero; therefore, there is no need to check for symmetry. The algorithm is similar to that of Nunes et al. (2005). However, the author performs extrema detection by comparing a candidate data point to its nearest-connected neighbors and suggests thin-plate splines, which are RBFs, for envelope construction.

5.1.5.1 Example

Figure 5.14 shows the use of the BEMD to remove shadows and clean the image before analysis. Figure 5.15 shows the basic subtraction of residue from the original image to obtain the final images. Figures 5.16 and 5.17 show how the BEMD can be used to preprocess track image (with corrosion) before the analysis like edge detection is obtained.

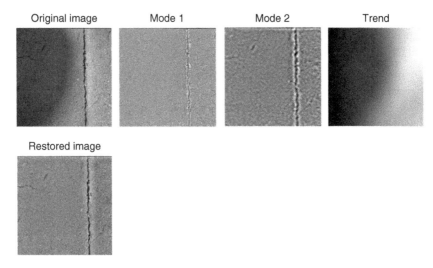

Figure 5.14 Using BEMD to remove shadows

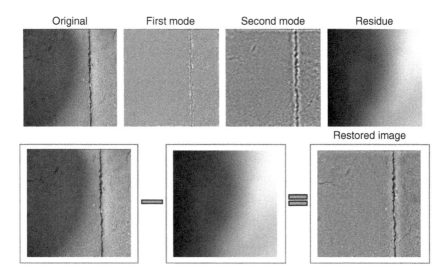

Figure 5.15 Subtraction of images

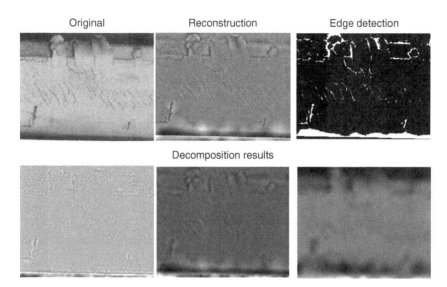

Figure 5.16 Preprocessing of track images

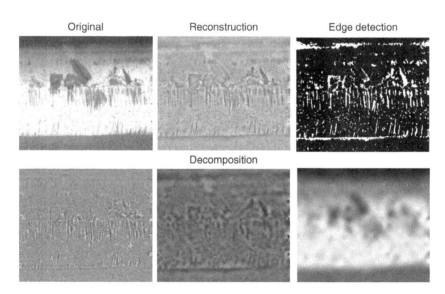

Figure 5.17 Preprocessing of track images

5.2 Axle Box Acceleration

5.2.1 General

Axle box acceleration (ABA) measurements measure the vibration of the wheel in the vehicle–track systems expected during wheel/rail interactions (Molodova et al., 2011). The axle box system consists of three main components as shown in Figure 5.18 (Oregui et al., 2016):

a) Accelerometers mounted on axle boxes to measure acceleration
b) A GPS system to measure the position of the vehicle
c) A piece of equipment (tachometer) to measure the speed

Therefore, the ABA is the vibration measured by the axle box supporting the axle. Also, the axle box is used to estimate the wheel loads and lateral forces by short wavelengths of track irregularities. In the vertical directions, the ABAs correlate strongly with wheel load; in the lateral directions, they correlate with the lateral forces (Tanaka, 2009).

Salvador et al. (2016) noted that axle box measurements offer a huge potential for obtaining more information about the track condition. The output signals can provide information on some singular track defects, such as squats, bolt tightness of fish-plated joints, and other short track defects. These can further be used to analyze rail corrugation, wheel/rail slippage, and in some cases the working conditions of turnouts. Table 5.1 shows types of defect and their wavelength ranges.

Generally, there are two main groups of defects: (a) ones associated with loss of track vertical geometry with wavelengths greater than 2 m and (b) those related to rail corrugation or isolated rail defects (squats, spalling).

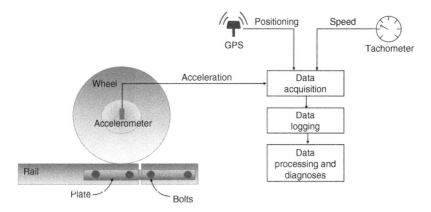

Figure 5.18 Schematic view of the axle box acceleration measuring and diagnosis system (Oregui et al., 2016). Reproduced with the permission of John Wiley and Sons

Table 5.1 Classification of track vertical defects upon their wavelengths.

Type of defect	Classification	Wavelength range (m)	Examples of defects
Rail corrugation and isolated rail defects	Very short	0.03–0.06	Rail joints, very short wavelength rail corrugation, small squats
	Short	0.06–0.25	Short wavelength rail corrugation, medium size squats
	Medium	0.25–0.60	Medium wavelength rail corrugation, large squats, turnout frogs
	Long	0.60–2	Long wavelength rail corrugation, ballast fouling
Loss of track vertical alignment	Short	2–25	Changes on track vertical stiffness
	Medium	25–70	Medium wavelength vertical misalignment
	Long	70–120	Long wavelength vertical misalignment

Salvador et al., 2016. Reproduced with the permission of Elsevier.

5.3 Analysis

Most of the analysis (signal outputs) are based on wavelets and Fourier analysis. Ayenu-Prah and Attoh-Okine (2010) compared the difference between Fourier, wavelet, and the HHT. Table 5.2 is the comparison between HHT, wavelet, and Fourier.

Fourier analysis involves signals with pure sinusoidal functions with constant amplitudes. Wavelet transform uses wavelets that can stretch or compress depending on whether the wavelet is on a low-frequency section or a

Table 5.2 Comparison between Fourier, wavelet, and HHT.

	Fourier	Wavelet	HHT
Basis	A priori	A priori	Adaptive
Frequency	Convolution: global, uncertainty	Convolution: regional, uncertainty	Differentiation: local, certainty
Presentation	Energy frequency	Energy time frequency	Energy time frequency
Nonlinear	Not easily defined	Not easily defined	Not easily defined
Nonstationary	No	Yes	Yes
Feature extraction	No	Yes	Yes
Theoretical base	Theory complete	Theory complete	Empirical

Table 5.3 Some applications of HHT in railway track engineering analysis.

Authors (Year)	Application	Source of data	Research aim
He et al. (2011)	Modal parameter identification of railway bridges	Nanjing Yangtze River Bridge	Identify modal parameters from monitoring vibrational data
Ho et al. (2010)	Rail structure analysis	Taoyuan Taiwan Railway; simulation data set	Explore the feasibility and applicability of using EMD and HHT on the analysis of track structure systems through laboratory testing and vibration testing
Li et al. (2016)	Railway wheel flat detection	Simulation data sets	Examine the axle box vibration response caused by wheel flats, considering the influence of both track irregularity and vehicle running speed on diagnosis results
Oukhellou et al. (2006)	Railway infrastructure system diagnosis (track circuit diagnosis)	French high-speed line (LGV), simulation data set	Detect its working state from one measurement signal that can be viewed as a superposition of several oscillations and periodic patterns called IMFs
Sun et al. (2012)	Fault diagnosis for railway track circuits trimming capacitors	Simulation data set	Ensure required dependability and availability levels of track circuit
Tsai et al. (2015)	Railway track inspection	Taiwan High-Speed Rail Corporation (THSRC)	Detect the rapid development of defects on railway systems
Zhao et al. (2013)	High-speed rail fault diagnosis	Simulation data set	Recognize fault patterns accurately and effectively
Zhu et al. (2015)	Railway line state detection signal processing (signal processing of track state detection)	Simulation data set	Accuracy of measurement, which is regularly influenced by the noise in the output signal of inertial measurement unit in the track state detecting process

high-frequency section of the signal. It has a finite duration and a zero mean. It suffers from the computation that arises out of defining a priori basis functions. A wavelet-based interpretation of a profile is only meaningful relative to the selected mother wavelet.

The use of EMD appears to be appropriate and effective of ABA signal output analysis. To analyze the signal of relevant components, the part of the signal is decomposed using the EMD approach and the noise that can be isolated from the signal varies. Furthermore, the HHT spectrum can be plotted to show the areas of high and low energy.

5.4   Remarks

Table 5.3 shows some applications of the HHT/EMD to railway track engineering. These appear to be only the initial applications. Wavelets/Fourier analysis have been used extensively, but this chapter presented the disadvantage of wavelets and Fourier. The use of the HHT/EMD is very promising and appropriate for railway track engineering signal and image processing. The HHT/EMD provides a more effective method of removing shadows in track images before any analysis. It is also capable of removing noise in the signals/images, which has not been implemented to the previous signal/image analyses. Hybrid HHT/EMD methods will provide more refined results for track signal and image analysis.

References

M. U. B. Altaf, T. Gautama, T. Tanaka, and D. P. Mandic. Rotation invariant complex empirical mode decomposition. In *2007 IEEE International Conference on Acoustics, Speech and Signal Processing - ICASSP '07*, pages III–1009–III–1012, IEEE, 2007.

A. Ayenu-Prah and N. Attoh-Okine. A criterion for selecting relevant intrinsic mode functions in empirical mode decomposition. *Advances in Adaptive Data Analysis*, 2(1):1–24, 2010. http://www.worldscientific.com/doi/abs/10.1142/S1793536910000367.

C. Damerval, S. Meignen, and V. Perrier. A fast algorithm for bidimensional EMD. *IEEE Signal Processing Letters*, 12(10):701–704, 2005. doi: 10.1109/LSP.2005.855548.

J.-J. Ding and S.-C. Lin. Ensemble Empirical Mode Decomposition, 2010.

J. S. Farnbach. The complex envelope in seismic signal analysis. *Bulletin of the Seismological Society of America*, 65(4):951–962, 1975.

X. H. He, X. G. Hua, Z. Q. Chen, and F. L. Huang. EMD-based random decrement technique for modal parameter identification of an existing railway bridge.

Engineering Structures, **33**(4):1348–1356, 2011. doi: 10.1016/j.engstruct .2011.01.012.

H.-H. Ho, P.-L. Chen, D. T.-T. Chang, and C.-H. Tseng. Rail structure analysis by empirical mode decomposition and Hilbert Huang transform. *Tamkang Journal of Science and Engineering*, **13**(3):267–279, 2010.

N. E. Huang, Z. Shen, S. R. Long, M. C. Wu, H. H. Shih, Q. Zheng, N.-C. Yen, C. C. Tung, and H. H. Liu. The empirical mode decomposition and the Hilbert spectrum for nonlinear and non-stationary time series analysis. *Proceedings of the Royal Society A: Mathematical, Physical and Engineering Sciences*, **454**(1971):903–995, 1998. doi: 10.1098/rspa.1998.0193.

C. Junsheng, Y. Dejie, and Y. Yu. Research on the intrinsic mode function (IMF) criterion in EMD method. *Mechanical Systems and Signal Processing*, **20**(4):817–824, 2006. doi: 10.1016/j.ymssp.2005.09.011.

D. M. Klionski, N. I. Oreshko, V. V. Geppener, and A. V. Vasiljev. Applications of empirical mode decomposition for processing nonstationary signals. *Pattern Recognition and Image Analysis*, **18**(3):390–399, 2008. doi: 10.1134/ S105466180803005X.

Y. Lei, Z. He, and Y. Zi. Application of the EEMD method to rotor fault diagnosis of rotating machinery. *Mechanical Systems and Signal Processing*, **23**(4):1327–1338, 2009. doi: 10.1016/j.ymssp.2008.11.005.

Y. Li, J. Liu, and Y. Wang. Railway wheel flat detection based on improved empirical mode decomposition. *Shock and Vibration*, **2016**:1–14, 2016.

A. Linderhed. Compression by image empirical mode decomposition. In *IEEE International Conference on Image Processing 2005*, volume 1, pages I–553. IEEE, 2005. doi: 10.1109/ICIP.2005.1529810.

M. Molodova, Z. Li, and R. Dollevoet. Axle box acceleration: measurement and simulation for detection of short track defects. *Wear*, **271**(1):349–356, 2011. doi: 10.1016/j.wear.2010.10.003.

J. C. Nunes, S. Guyot, and E. Delechelle. Texture analysis based on local analysis of the Bidimensional Empirical Mode Decomposition. *Machine Vision and Applications*, **16**(3):177–188, 2005. doi: 10.1007/s00138-004-0170-5.

M. Oregui, S. Li, A. Núñez, Z. Li, R. Carroll, and R. Dollevoet. Monitoring bolt tightness of rail joints using axle box acceleration measurements. *Structural Control and Health Monitoring*, 2016. doi: 10.1002/stc.1848.

L. Oukhellou, P. Aknin, and E. Delechelle. Railway infrastructure system diagnosis using empirical mode decomposition and Hilbert transform. In *2006 IEEE International Conference on Acoustics Speed and Signal Processing Proceedings*, **3**(1), pages III–1164–III–1167, 2006. doi: 10.1109/ICASSP.2006.1660866. http://ieeexplore.ieee.org/xpls/abs_all.jsp?arnumber=1660866/nhttp:// ieeexplore.iccc.org/lpdocs/epic03/wrapper.htm?arnumber=1660866.

G. Rilling and P. Flandrin. One or two frequencies? The empirical mode decomposition answers. *IEEE Transactions on Signal Processing*, **56**(1):85–95, 2008. doi: 10.1109/TSP.2007.906771.

J. Rudi. *Empirical mode decomposition via adaptiver wavelet-approximation*. PhD thesis, Universitat Paderborn, 2010. https://pdfs.semanticscholar.org/876b/5c96e19cc9611f5d6536921ba59e643be0e5.pdf.

P. Salvador, V. Naranjo, R. Insa, and P. Teixeira. Axlebox accelerations: their acquisition and time–frequency characterisation for railway track monitoring purposes. *Measurement*, **82**:301–312, 2016. doi: 10.1016/j.measurement.2016.01.012.

S. P. Sun, H. B. Zhao, and G. Zhou. A fault diagnosis approach for railway track circuits trimming capacitors using EMD and Teager energy operator. *WIT Transactions on the Built Environment*, **127**:167–176, 2012. doi: 10.2495/CR120151.

H. Tanaka, A Method of Managing Wheel Loads and Lateral Forces Using Axle-Box Acceleration, Railway Technology Avalanche, no. 27, 2009.

T. Tanaka and D. P. Mandic. Complex empirical mode decomposition. *IEEE Signal Processing Letters*, **14**(2):101–104, 2007. doi: 10.1109/LSP.2006.882107.

C.-W. Tsai, C.-F. Lai, H.-C. Chao, and A. V. Vasilakos. Big data analytics: a survey. *Journal of Big Data*, **2**(1):21, 2015. doi: 10.1186/s40537-015-0030-3.

Z. Wu and N. E. Huang. Ensemble empirical mode decomposition: a noise-assisted data analysis method. *Advances in Adaptive Data Analysis*, **1**(1):1–41, 2009. doi: 10.1142/S1793536909000047.

L.-H. Zhao, C.-L. Zhang, K.-M. Qiu, and Q.-L. Li. A fault diagnosis method for the tuning area of jointless track circuits based on a neural network. *Proceedings of the Institution of Mechanical Engineers Part F-Journal of Rail and Rapid Transit*, **227**(4):333–343, 2013. doi: 10.1177/0954409713480453.

F. Zhou, M. Xing, X. Bai, G. Sun, and Z. Bao. Narrow-band interference suppression for SAR based on complex empirical mode decomposition. *IEEE Geoscience and Remote Sensing Letters*, **6**(3):423–427, 2009. doi: 10.1109/LGRS.2009.2015340.

W.-F. Zhu, H.-Z. Ma, X.-D. Chai, and S.-B. Zhen. The research of railway line state detection signal processing method based on EMD. *Open Journal of Safety Science and Technology*, **5**:63–68, 2015.

6

Tensors – Big Data in Multidimensional Settings

6.1 Introduction

Railway track monitoring data is usually high dimensional. The structure of this high-dimensional data in most cases and conditions can be characterized by a relatively small number of parameters. Reducing the data dimensionality presents the engineer with different opportunities, including visualization of the intrinsic structure of the data and more efficient data for developing the appropriate models, such as prediction. The "flat-world view" of 2-way matrix applications may be insufficient in making inferences in interdependent infrastructures. In the current literature on infrastructure data analysis, most high-dimensional data are inappropriately represented, making it very difficult to develop the correct models for further analysis.

For example, the use of image analysis in infrastructure monitoring requires a new form of data representation in large-scale civil infrastructure systems. In general, the resilience of an interdependent network can be presented as multiple graphs, and the adjacency tensor can provide the framework for addressing the resilience of interdependent networks.

Tensors appear to be an appropriate way to represent high-dimensional data and their interdependences in railway track infrastructure (Attoh-Okine, 2016). Tensor factorization and decomposition are becoming major tools for large multidimensional data analysis. Factorizing tensors has better advantages than traditional matrix factorization, such as the uniqueness of the optimal solution, and the decomposition can explicitly account for the multiway structure of the data (Mørup, 2011). The application of the tensor, apart from addressing the previous shortcomings, will provide a platform for performing data mining applications. Sun et al. (2006) noted that the tensor approach is capable of detecting anomalies in data. The anomaly detection can proceed from the broadest level to a more specific level. Sun et al. (2006) discuss the process of using tensors in computer network modeling, which has some similarities with

Big Data and Differential Privacy: Analysis Strategies for Railway Track Engineering, First Edition. Nii O. Attoh-Okine.
© 2017 John Wiley & Sons, Inc. Published 2017 by John Wiley & Sons, Inc.

large civil network infrastructure. This property is very important, especially in analyzing the efficient performance of large-scale infrastructure systems. The approach can further determine a quantitatively capable way of finding the dimension of the error nodes. Some of the advantages of tensor analysis include ((Barnathan, 2010); (Acar et al., 2007)):

- Full exploitation of the high-order track infrastructure data – the tensor can represent the high-dimensional data sets without losing any information.
- Excellent representation for spatiotemporal data.
- Tensor techniques are capable of making inferences across different infrastructures, for example, when there is interdependence.
- Tensors are very efficient in handling missing data due to loss of information and errors during the data collection process.

Some of the major disadvantages of the tensor applications are that the storage requirements tend to increase exponentially in scale, the methods tend to be global, and it fails to take spatiotemporal locality into account. Depending on the objective of the analysis, this situation can be advantageous, especially when looking at network or project analysis. Low-order analyses like the matrix have been long-established techniques in infrastructure analysis; therefore, one has to put extra effort into explaining the output to regular decision makers.

6.2   Notations and Definitions

Kolda and Bader (2009) defined a tensor as a multidimensional array and formally introduced the explanation to differentiate the definition from tensors in physics and engineering, which are generally referred to as tensor fields in mathematics. Also, "tensor" is a multilinear algebra term, which generalizes the concepts of "vectors" and "matrices." A vector is a 1-order tensor, matrix data is a 2-order tensor, and a 3-order tensor is a cube-like data structure. Any data structure greater than 3-order becomes an abstract data structure that is a generalization of vectors and matrices (Liu et al., 2005).

A tensor is a multiway array of data and is denoted by bold capital letters, for example, $Y \in \mathbb{R}^{I_1 \times I_2 \times \cdots, I_N}$ is an N-way tensor whose entries are denoted by one of the following: $Y(i_1, i_2, \ldots, i_N) = y i_1 i_2 \ldots, i_N$ (Cichocki et al., 2009). The order of tensors is the number of dimensions, also called modes. Matrices (2-way tensors) are denoted by uppercase letters, $Y \in \mathbb{R}^{I \times J}$ for three-dimensional (3D) tensors, $Y \in \mathbb{R}^{I \times J \times K}$, their frontal slices, lateral slices, and horizontal slices are denoted, respectively, by $Y_K = Y_{::k}, Y_{:j:}$, and $Y_{i::}$. The n mode fiber is obtained by fixing all indices except the index in one dimension. For a 3-way (3D) tensor,

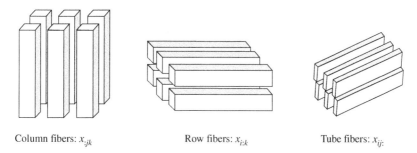

Column fibers: $x_{:jk}$ Row fibers: $x_{i:k}$ Tube fibers: $x_{ij:}$

Figure 6.1 3D tensor fibers

the n-mode fibers are called column fibers ($n = 1$), row fibers ($n = 2$), and tube fibers ($n = 3$) ((Caiafa and Cichocki, 2010); (Phan and Cichocki, 2009)). A tube (vector) at position (i, j) along the mode-3 is denoted by $\mathbf{y}_{ij:}$, and the corresponding tubes along the mode-2 and mode-1 are $y_{i:k}$ and $y_{:jk}$, respectively. Fibers are the higher-order equivalent of matrix rows and columns and are defined by fixing all but one of the indices, as shown in Figure 6.1. Slices are two-dimensional (2D) sections of a tensor and are defined by fixing all but two indices. Figure 6.2 shows the horizontal, lateral, and frontal slices of a 3-order tensor (Kolda and Bader, 2009).

Apart from the basic definitions, terminologies and brief explanations are needed of the following: (a) Kronecker, (b) the Khatri–Rao, (c) Hadamard products, and (d) element-wise division is denoted as follows: $\otimes, \odot, \circledast, and\ \varnothing$. The Kronecker product of matrices $A \in \mathbb{R}^{I \times J}$ and $B \in \mathbb{R}^{K \times L}$ is denoted by $A \otimes B$; the result is a matrix $(IK) \times (JL)$, defined by

$$A \otimes B = \begin{bmatrix} a11B & a12B & \dots & a1JB \\ a21B & a22B & \dots & a2JB \\ \vdots & \vdots & & \vdots \\ aI1B & aI2B & \dots & aIJB \end{bmatrix},$$

$$A \otimes B = [a_1 \otimes b_1 \quad a_1 \otimes b_2 \quad a_1 \otimes b_3 \quad a_J \otimes b_{L-1} \quad a_J \otimes b_L] = R^{IJ \times JL}.$$

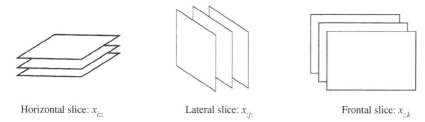

Horizontal slice: $x_{i::}$ Lateral slice: $x_{:j:}$ Frontal slice: $x_{::k}$

Figure 6.2 3D tensor slices

For any three matrices A, B, and C, where B and C have the same size, the Kronecker product satisfies these properties:

$$(A \otimes B)^T = A^T \otimes B^T \tag{6.1}$$

$$A \otimes (B + C) = (A \otimes B) + (A \otimes C) \tag{6.2}$$

$$(B + C) \otimes A = (B \otimes A) + (C \otimes A) \tag{6.3}$$

$$(A + B) \otimes (C + D) = AC \otimes BD \tag{6.4}$$

$$c(A \otimes B) = (cA) \otimes B = A \otimes (cB). \tag{6.5}$$

The Khatri–Rao product is the matching column-wise Kronecker product. Given the matrices $A \in \mathbb{R}^{I \times J}$ and $B \in \mathbb{R}^{K \times L}$, their Khatri–Rao product is denoted by $A \odot B$ with the size $(IJ) \times K$; $A \odot B = [a_1 \otimes b_1 \ a_2 \otimes b_2 \ a_3 \otimes b_3 \ a_K \otimes b_K]. = C^{IK \times J}$. When a and b are vectors, the Khatri–Rao and Kronecker products are identical.

The Khatri–Rao product has the following properties:

- Associative

$$A \odot (B \odot C) = (A \odot B) \odot C \tag{6.6}$$

- Distributive

$$(A + B) \odot C = A \odot C + B \odot C \tag{6.7}$$

- Non-commutative

$$A \odot B \neq B \odot A \tag{6.8}$$

The Hadamard product is the element-wise matrix product; given the matrices $A \in \mathbb{R}^{I \times J}$ and $B \in \mathbb{R}^{I \times J}$ (note that both are size $I \times J$), the Hadamard product is as follows: $A * B = \begin{bmatrix} a_{11}b_{11} & a_{12}b_{12} & \cdots & a_{1J}b_{iJ} \\ a_{21}b_{21} & a_{22}b_{22} & \cdots & a_{2J}b_{2J} \\ \vdots & \vdots & & \vdots \\ a_{I1}b_{I1} & aI_2b_{I2} & \cdots & a_{IJ}b_{IJ} \end{bmatrix}$.

Tensors can be multiplied together. An example of an n-node product, tensor by a matrix, or a vector in mode n is as follows: $Y \in \mathbb{R}^{I_1 \times I_2 \times \cdots I_N}$ multiplied with matrix $K \in \mathbb{R}^{J \times I_n}$ is denoted by $Y x_n K$ and is the size of $I_1 \times \cdots I_{n-1} \times J \times I_{n+1} \times \cdots \times I_N$. Each mode-$n$ fiber is multiplied by the matrix K. The n-mode vector product of Y with a vector $w \in \mathbb{R}^{I_n}$ is denoted by $Y \bar{x}_n w$. The final result is of order $N - 1$. The size is $I_1 \times \cdots \times I_{n-1} \times I_{n-1} \times \cdots \times I_N$. In mode-$n$ multiplication, precedence is very important (Kolda and Bader, 2009).

The nth mode matricizing and unmatricizing operations map a tensor into a matrix and vice versa. Matricization is sometimes referred to as unfolding and flattening. For example, a $2 \times 3 \times 5$ tensor data can be arranged into a 6×5 or 3×10 matrix. Kolda (2006) presented an approach in the treatment of matricization. The n-mode matricization of tensor $(\underline{Y} \in \mathbb{R}^{I_1 \times I_2 \times \dots I_N})$ is denoted by a matrix $Y_{(n)}$:

$$\underline{Y} \in \mathbb{R}^{I_1 \times I_2 \times \dots I_N} \longrightarrow Y_{(n)}^{I_1 \times I_2 \times \dots I_N}. \tag{6.9}$$

Unmatricizing is as follows:

$$Y_{(n)}^{I_1 \times I_2 \times \dots I_N} \longrightarrow \underline{Y} \in .\mathbb{R}^{I_1 \times I_2 \times \dots I_N} \tag{6.10}$$

In some instances, it is more appropriate to represent tensors and matrices as vectors; this process is referred to as vectorization. For example, if the matrix $Y = [y_1, y_2, \dots, y_T] \in \mathbb{R}^{I \times T}$, the vectorization is defined as

$$y = \text{vec}(Y) = [y_1, y_2, \dots, y_T] \in .\mathbb{R}^{I \times T} \tag{6.11}$$

The vec-operator applied on matrix Y stacks its column into a vector. The vectorization of the third order tensor, $\underline{Y} \in \mathbb{R}^{I_1 \times I_2 \times I_3}$, can be expressed in the following form:

$$\text{vec}(\underline{Y}) = \text{vec}(Y_{(1)}) = \left[\text{vec}(Y_{(:,:1)})^T, \text{vec}(Y_{(:,:2)})^T, \dots, \text{vec}(Y_{(:,:3)})^T \right]^T \in .\mathbb{R}^{I_1 \times I_2 \times I_3} \tag{6.12}$$

The basic properties of vec-operators are:

$$\text{vec}(cA) = c\text{vec}(A) \tag{6.13}$$

$$\text{vec}(A + B) = \text{vec}(A) + \text{vec}(B) \tag{6.14}$$

$$\text{vec}(A)^T \text{vec}(B) = \text{trace}(A^T B) \tag{6.15}$$

$$\text{vec}(ABC) = (C^T \otimes A)\text{vec}(B). \tag{6.16}$$

6.3   Tensor Decomposition Models

The two most widely used tensor decomposition models are the Tucker model and the canonical decomposition (CANDECOMP) (which is also known as CP or parallel factor analysis (PARAFAC)) model. The Tucker model can be described as the decomposition of Nth order tensor $\underline{Y} \in \mathbb{R}^{I_1 \times I_2 \times \dots I_N}$ into an unknown core tensor. $\underline{G} \in \mathbb{R}^{J_1} \times J_2 \dots \times J_N$ is multiplied by a set of N unknown component matrices, and $A(n) = [a_1^{(n)}, a_2^{(n)}, \dots, a_{J_n}^{(n)}] \in \mathbb{R}^{I_n \times J_n} n = 1, 2, \dots, N)$ represents the factor loadings (Cichocki et al., 2009):

$$\underline{Y} = \sum_{j_1=1}^{J_1} \sum_{j_2=1}^{J_2} \dots \sum_{j_N=1}^{J_N} g_{j_1 j_2 \dots j_N} (a_{j_1}^{(1)} \circ a_{j_2}^{(2)} \circ \dots a_{j_N}^{(N)}) + \underline{E} \tag{6.17}$$

$$\underline{Y} = \underline{G}x_1 A^{(1)} x_2 A^{(2)} \dots x_N A^{(N)} + \underline{E} \tag{6.18}$$

$$\underline{Y} = \underline{G}x\{A\} + \underline{E} = \underline{Y}^* + \underline{E}. \tag{6.19}$$

E is the error tensor, and ∘ is the outer product. Mørup (2011) and Phan and Cichocki (2009) presented methods of analysis and figures to demonstrate the 3-way Tucker model. The Tucker model is not unique.

The Tucker decomposition of a 3D matrix $\mathbf{T} = \{t_{ijk}\}$, $i = 1, \dots, I$, $j = 1, \dots$, J, $k = 1, \dots, K$ decomposes it into a small 3D matrix $\mathbf{G} = \{g_{pqr}\}$, $p = 1, \dots$, P, $q = 1, \dots, Q$, $r = 1, \dots, R$ and three 2D matrices $\mathbf{X} = \{x_{ip}\}$, $\mathbf{Y} = \{y_{jq}\}$, and $\mathbf{Z} = \{z_{kr}\}$. Based on this representation, a tensor element t_{ijk} can be approximated by

$$t_{ijk} \approx \sum_{p=1}^{P} \sum_{q=1}^{Q} \sum_{r=1}^{R} g_{pqr} x_{ip} y_{jq} z_{kr}. \tag{6.20}$$

PARAFAC is a special case of the Tucker model in which the core tensor is a cubical super diagonal or super-identity tensor with nonzero elements only on the super diagonal ($J_1 = J_2 = \cdots = J_N$). Therefore, the decomposition factorizes a tensor into a sum of component rank-one tensors. If the tensor $\underline{Y} \in \mathbb{R}^{I_1 \times I_2 \times I_3}$, then

$$\underline{Y} \approx \sum_{r=1}^{R=1} a_r \circ b_r \circ c_r + E. \tag{6.21}$$

R is a positive integer, and $a_r \in \mathbb{R}^{I_1}$, $b_r \in \mathbb{R}^{I_2}$, and $c_r \in \mathbb{R}^{I_3}$ for $r = 1, \dots, R$. The Tucker model allows for extraction of different numbers of factors in each mode and permits interaction, while PARAFAC does not have those capabilities. The Tucker model advantages are disadvantages in some situations. For example, the Tucker method has a high ability to compress data with the minimum number of components. Unfortunately, each component may have different interactions. Acar and Yener (2009) presented different forms of the Tucker, PARAFAC, and another alternative model; they presented different algorithms for the solution of different problems.

6.3.1 Nonnegative Tensor Factorization

Nonnegative matrix factorization (NMF) determines the recovery of hidden structures, trends, or patterns from redundant data (Cichocki et al., 2009). Therefore, the NMF problem can be described as follows:

Given a nonnegative matrix Y of size $I \times T$, find two nonnegative matrices A (size $I \times J$) and X (size $J \times T$) such that their product AX is approximately equal to Y.

The elements of all A are ≥ 0; J is selected to be much smaller than I and T. Mathematically, it can be expressed as follows

$$Y = AX + E, \tag{6.22}$$

where E represents noise or error. Various algorithms and measures are suggested in the literature (Ho, 2008).

Schlink and Thiem (2010) present the following. Given a set of m multivariate n-dimensional data vectors, placed in the columns of a nonnegative $n \times m$ matrix V with $V_{ij} \geq 0$, $\forall i,j$, where m is the number of examples in the data set (observations), and a natural number $r \leq \min(n, m)$, NMF tries to approximately factorize the data matrix into nonnegative matrices $W \in R^{n \times r}$ and $H \in R^{r \times m}$ such that

$$V \approx WH, \; i = 1, \ldots, n; \; j = 1, \ldots, m, \tag{6.23}$$

$$W_{ia} \geq 0, \; H_{bj} \geq 0 \;\; a = 1, \ldots, r; \;\; b = 1, \ldots, r.$$

Each column V represents an object, for example, the set of pixel values of an image. Each row corresponds to a certain variable, for example, the position of one of the pixels. Alternatively, Equation 6.23 can be rewritten column-wise, that is, $V_{\bullet j} \approx W * H_{\bullet j}$, which means that each data vector in V is approximated by a linear combination of the columns of W weighted by the components of $H_{\bullet j} \cdot W$ contains characteristic basis patterns.

One way to find the matrix factors W and H is to minimize the difference between V and WH in terms of the squared Euclidean distance:

$$f(W, H) = \frac{1}{2} \sum_{i=1}^{n} \sum_{j=1}^{m} (V_{ij} - (WH)_{ij})^2 = \frac{1}{2} \|V - WH\|_F^2 \to \min_{W,H}, \tag{6.24}$$

$$W_{ia} \geq 0, H_{bj} \geq 0, \forall \, i, a, b, j,$$

where $\|\bullet\|$ denotes the Frobenius norm. Another way is the minimization of the cost function

$$g(W, H) = \sum_{i=1}^{n} \sum_{j=1}^{m} \left(V_{ij} \log \frac{V_{ij}}{(WH)_{ij}} - V_{ij} + (WH)_{ij} \right), \tag{6.25}$$

which is the (generalized) Kullback–Leibler divergence (subject to the same nonnegativity constraints as in Equation 6.24) (Schlink and Thiem, 2010).

Lee and Seung (1999) formally define NMF as follows, given a nonnegative, real-valued data matrix $V \in \mathbb{R}_+^{n \times m}$ such that

$$V \approx WH = \sum_{i=1}^{k} w_i h_i^T \triangleq \sum_{i=1}^{k} w_i \circ h_i, \tag{6.26}$$

where w_i are the columns of W, h_i^T are rows of H, and W is the basis vector; finally, V forms a linear combination of the column vectors W. Selection of k satisfies the following condition: $(n + m)k < nm$. This will result in a compressed version of the original data matrix (Benetos and Kotropoulos, 2010).

In a noisy situation, the approximation of a nonnegative matrix can be expressed as follows:

$$\min_{W \geq 0, H \geq 0} \|V - WH^T\| = \min_{w_i \geq 0, h_i \geq 0} \|V - \sum_{i=1}^{k} w_i \circ h_i\|. \tag{6.27}$$

The generalization of a NMF to tensors of a higher order leads to nonnegative PARAFAC models; this process is sometimes referred to as nonnegative tensor factorization (NTF) (Lim and Comon, 2009).

NTF can be formally defined as follows. Given a nonnegative tensor $\tau \in \mathbb{R}^{m \times n \times P}$ and a positive integer k, find the nonnegative vectors $x^{(i)} \in \mathbb{R}^{m \times 1}$, $y^{(i)} \in \mathbb{R}^{n \times 1}$, and $z^{(i)} \in \mathbb{R}^{P \times 1}$ to minimize the functional $\frac{1}{2} \left\| \tau - \sum_{k=1}^{K} x^{(i)} \circ y^{(i)} \circ z^{(i)} \right\|_F^2$.
This determines three matrices $X^{(i)} \in \mathbb{R}^{m \times k}$, $Y^{(i)} \in \mathbb{R}^{n \times k}$, and $Z^{(i)} \in \mathbb{R}^{p \times k}$.

More generally, NTF can be expressed as follows: given an Nth order tensor $\underline{Y} \in \mathbb{R}^{I_1 \times I_2 \times \cdots \times I_N}$ and a positive integer J, factorize $\underline{Y}$ into a set of N nonnegative component matrices $A^{(n)} = [a_1^{(n)}, a_i^{(n)}, \ldots, a_j^{(n)}] \in \mathbb{R}^{I_n \times J}$ $(n = 1, 2, \ldots, N)$ representing the loading features $\underline{Y} = \sum_{j=1}^{J} a_j^{(1)} \circ a_2^{(2)} \circ \ldots \circ a_j^{(N)} + \underline{E}$.

The aim of nonnegative tensor is to extract the data-dependent nonnegative basis function, and the target data can be expressed by the linear and non-linear combination of the nonnegative components (Liu et al., 2011). NTF is the extension of NMF.

NTF is the generalization of nonnegative matrix factorization. It lends itself to computing a nonnegative low-rank approximation to a multiway data array (Schmidt and Mohamed, 2009). Also, it recovers hidden nonnegative common structures or patterns from the data (Cichocki et al., 2008). The approximation of factorization leads to the solution of the optimization problem [...... so many authors]. The main idea, therefore, in NTF is the decomposition of an N-way array $(y \ldots 0)$ to the sum of Krank $- 1$ tensors that are outer products of nonnegative vectors $Y \in \mathbb{R}_+^{I_1 \times I_2 \times \ldots I_N}$:

$$Y = \sum_{j=1}^{k} u_j^{(1)} \circ u_j^{(2)} \circ \ldots \circ u_j^{(n)}, \tag{6.28}$$

where $u_j^{(i)} \in \mathbb{R}_+^{I_i}$ and $j = 1, 2, \ldots, k$.

6.4 Application

Figure 6.3 is an example of cross-level measurement at different dates. Figure 6.4 shows combinations of different geometry defect parameters at a selected time. A major analytical issue is how to use the multidimensional data in both modeling and maintenance decision-making. Figure 6.5 shows the data structure of the cross-level used for the analysis.

Figure 6.7 shows the correlation analysis of the cross-level data. Figure 6.7 demonstrates that the cross-level exhibit non-normal distribution. A simple correlation plot showed August and July to be similar. We can justify the use of

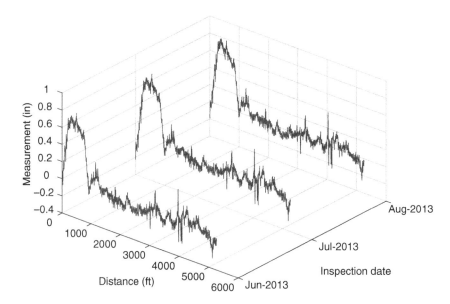

Figure 6.3 Cross-level measurement at different dates

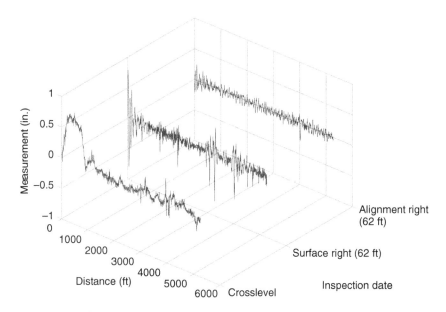

Figure 6.4 Track geometry parameters measurements

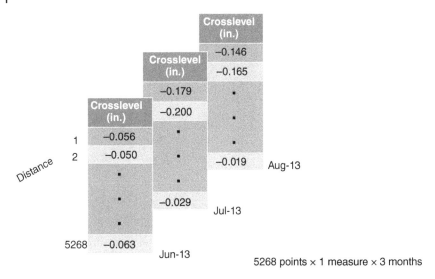

Figure 6.5 Data structure for the cross-level

tensor factorization with the introduction of additional months. That is when it may be difficult to analyze with traditional methods. For Figure 6.6, the loading plot for the cross-level shows the June measurement. Figure 6.8 shows the loading plots for distance points divided into four quadrants.

Figure 6.9 is the data structure for cross-level surface (right) and alignment on the same data. Figure 6.10 shows the correlation structure of cross-level, surface (right), and alignment. It appears that the surface (right) and alignment

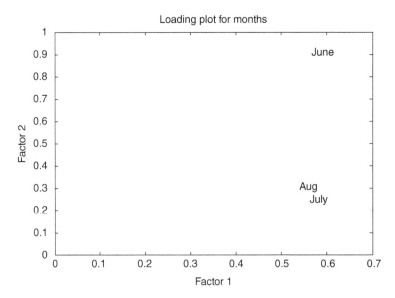

Figure 6.6 Loading plot for cross-level

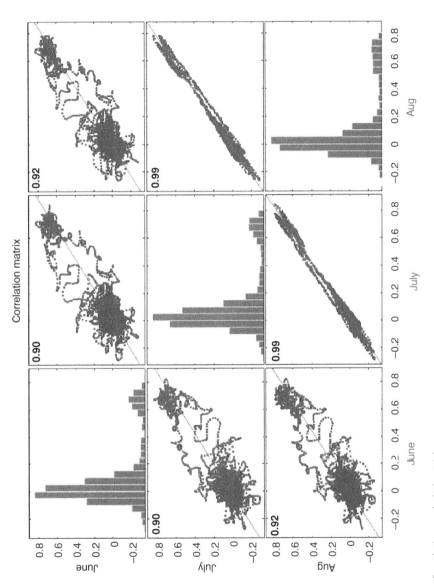

Figure 6.7 Correlation analysis (matrix)

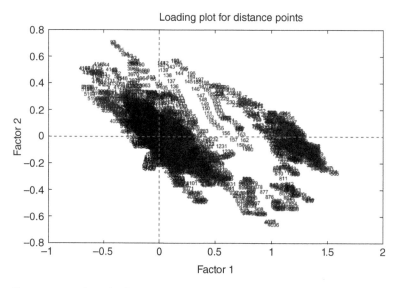

Figure 6.8 Loading plot for distance points

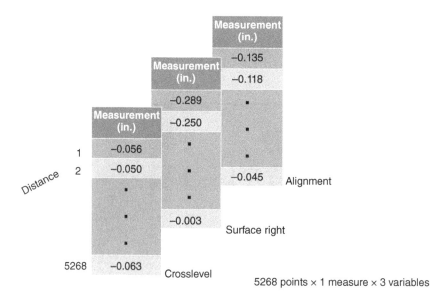

Figure 6.9 Data structure for cross-level, surface (right) and alignment on same date

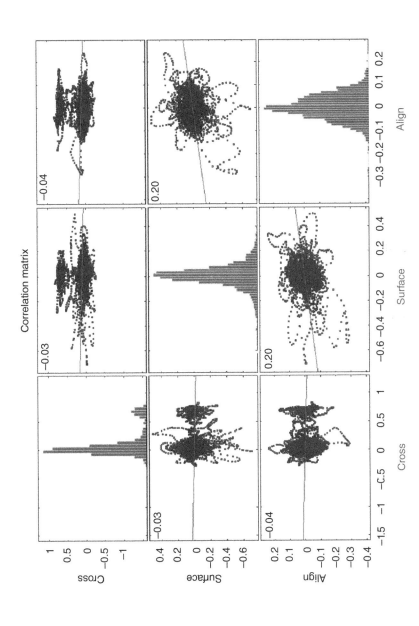

Figure 6.10 Correlation matrix

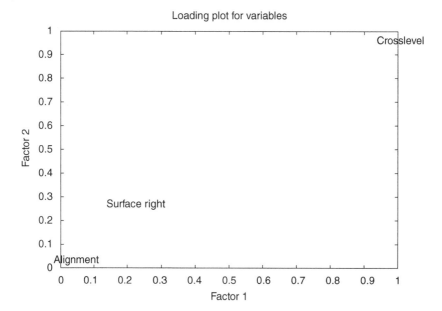

Figure 6.11 Loading plots

are normally distributed. Figure 6.11 is the loading plot. It shows the inputs (based on the three variables –cross-level, surface (right), and alignment).

Figure 6.12 shows that the majority of the points are located in the upper left and lower right quadrants. Although the factors were correlated, the plots did not show any interesting trends.

6.5 Remarks

To effectively interpret rail geometry data in a multidimensional approach, the use of 2D analysis of railway data fails to address the proper influence of different parameters in time after maintenance and the interaction of different variables in time. The current example presents a basic overview of tensor decomposition as a tool to address multidimensional inputs of track data for further analysis. The correct formulation of the problem is the key in gaining more in-depth information about the interaction of the different variables with the behavior of a related variable in a time span.

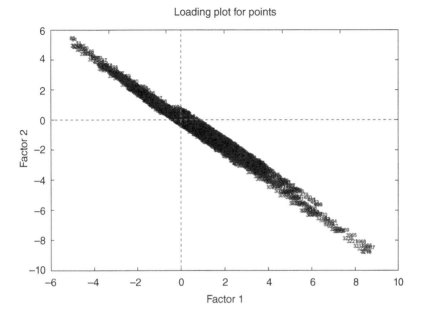

Figure 6.12 Loading plots for points

References

E. Acar, C. Aykut-Bingol, H. Bingol, R. Bro, and B. Yener. Multiway analysis of epilepsy tensors. *Bioinformatics*, **23**(13):10–18, 2007. doi: 10.1093/bioinformatics/btm210.

E. Acar and B. Yener. Unsupervised multiway data analysis: a literature survey. *IEEE Transactions on Knowledge and Data Engineering*, **21**(1):6–20, 2009. doi: 10.1109/TKDE.2008.112.

N. Attoh-Okine. *Resilience Engineering: Models and Analysis.* Cambridge University Press, 2016. http://www.cambridge.org/us/academic/subjects/engineering/engineering-mathematics-and-programming/resilience-engineering-models-and-analysis?format=HB&isbn=9780521193498; https://books.google.com/books?id=O_-lCwAAQBAJ&pg=PR1&lpg=PR1&dq=Resilience+engineering:+models+and+analysis+attoh-okine&source=bl&ots=pd9zBxkGAv&sig=55K7-scwsEVae1oN5ZvBwTH4vNk&hl=en&sa=X&ved=0ahUKEwiA87rs0djRAhVG4iYKHWZrDh0Q6AEIYDAH#v=onepage&q=Resilience%20engineering%3A%20models%20and%20analysis%20attoh-okine&f=false.

M. Barnathan. *Mining complex high-order datasets*. PhD thesis, 2010. http://dl
.acm.org/citation.cfm?id=1970509.

E. Benetos and C. Kotropoulos. Non-negative tensor factorization applied to
music genre classification. *IEEE Transactions on Audio, Speech, and Language
Processing*, **18**(8):1955–1967, 2010. doi: 10.1109/TASL.2010.2040784.

C. F. Caiafa and A. Cichocki. Generalizing the column-row matrix decomposition
to multi-way arrays. *Linear Algebra and its Applications*, **433**(3):557–573, 2010.
doi: 10.1016/j.laa.2010.03.020.

A. Cichocki, R. Zdunek, and S.-I. Amari. Nonnegative matrix and tensor
factorization. *IEEE Signal Processing Magazine*, **25**(1):142–145, 2008.
doi: 10.1109/MSP.2007.911394.

A. Cichocki, R. Zdunek, A. H. Phan, and S.-I. Amari. *Nonnegative Matrix and
Tensor Factorizations: Applications to Exploratory Multi-Way Data Analysis
and Blind Source Separation*. Wiley Publishing, 2009. ISBN: 0470746661,
9780470746660. http://www.researchgate.net/publication/237145400_
Nonnegative_Matrix_and_Tensor_Factorizations_–_Applications_to_
Exploratory_Multi-way_Data_Analysis_and_Blind_Source_Separation_–_
Chapters_1_and_2 http://dl.acm.org/citation.cfm?id=1822971.

N.-D. Ho. Nonnegative matrix factorization algorithms and applications. PhD
Thesis, Université Catholique de Louvain, 2008.

T. G. Kolda. Multilinear operators for higher-order decompositions. Technical
report, 2006. http://www.ca.sandia.gov/~tgkolda/pubs/pubfiles/SAND2006-
2081.pdf; http://citeseerx.ist.psu.edu/viewdoc/summary?doi=10.1.1.125.2423.

T. G. Kolda and B. W. Bader. Tensor decompositions and applications. *SIAM
Review*, **51**(3):455–500, 2009. http://www.sandia.gov/~tgkolda/pubs/pubfiles/
TensorReview.pdf; http://epubs.siam.org/doi/abs/10.1137/07070111X.

D. D. Lee and H. S. Seung. Learning the parts of objects by non-negative matrix
factorization. *Nature*, **401**(6755):788–91, 1999. doi: 10.1038/44565.

L.-H. Lim and P. Comon. Nonnegative approximations of nonnegative tensors.
Journal of Chemometrics, **23**(1):432–441, 2009. http://onlinelibrary.wiley.com/
doi/10.1002/cem.1244/abstract.

J. Liu, J. Liu, P. Wonka, and J. Ye. Sparse non-negative tensor factorization using
columnwise coordinate descent. *Pattern Recognition*, **45**:649–656, 2011.
doi: 10.1016/j.patcog.2011.05.015.

N. Liu, B. Zhang, J. Yan, Z. Chen, W. Liu, F. Bai, and L. Chien. Text representation:
from vector to tensor. In *5th IEEE International Conference on Data Mining
(ICDM'05)*, pages 725–728. IEEE, 2005. ISBN: 0-7695-2278-5. http://ieeexplore
.ieee.org/lpdocs/epic03/wrapper.htm?arnumber=1565767.

M. Mørup. Applications of tensor (multiway array) factorizations and
decompositions in data mining. *Wiley Interdisciplinary Reviews: Data Mining
and Knowledge Discovery*, **1**(1):24–40, 2011. doi: 10.1002/widm.1.

A. H. Phan and A. Cichocki. Analysis of interactions among hidden components
for Tucker model. In *APSIPA Annual Summit and Conference*, 2009. http://

www.researchgate.net/publication/39999866_Analysis_of_Interactions_
Among_Hidden_Components_for_Tucker_Model.

U. Schlink and A. Thiem. Non-negative matrix factorization for the identification
of patterns of atmospheric pressure and geopotential for the Northern
Hemisphere. *International Journal of Climatology*, **30**(6):909–925, 2010.
doi: 10.1002/joc.1942.

M. N. Schmidt and S. Mohamed. Probabilistic non-negative tensor factorization
using Markov chain Monte Carlo, 2009.

J. Sun, D. Tao, and C. Faloutsos. Beyond streams and graphs. In *Proceedings of the
12th ACM SIGKDD International Conference on Knowledge Discovery and
Data Mining - KDD '06*, page 374, ACM Press, New York, 2006.
ISBN: 1595933395. http://dl.acm.org/citation.cfm?id=1150402.1150445.

7

Copula Models

7.1 Introduction

Almost all the analysis in track geometry modeling is based on the assumption that all variables are normally distributed. This assumption may not always be valid. Track degradation modeling is often multidimensional; therefore, analysis usually involves modeling several random variables (Attoh-Okine, 2013). In previous studies, the multivariate normal distribution is frequently assumed. The multivariate normal assumption dictates that the type of association between the normal margins be linear (Vaz de Melo Mendes and Martins de Souza, 2004).

In track applications, non-normality occurs in different ways (Yan, 2005): (a) the marginal distributions of some variables may not be normal and (b) in some cases, if all the marginal distributions are normal, jointly these variables may not be multivariate normal. The appropriateness of the multivariate normal distribution and correlation measure of association, as they have been discussed in the literature, has forced engineers to consider the following practical issues and questions (Vaz de Melo Mendes and Martins de Souza, 2004):

- What are the appropriate multivariate distributions for modeling infrastructure data?
- Which dependence measures should be computed to appropriately explain the several types of associations?
- What are the effects of the assumed dependence structure on estimates of risk measures?

This implies that the individual behavior of rail track condition variables or their transformation must be characterized by one and only one parametric family of univariate distributions. Multivariate distributions have been dominated by normal distributions because multivariate normal distributions have manageable characteristics, including their marginal distributions, which are

Big Data and Differential Privacy: Analysis Strategies for Railway Track Engineering, First Edition. Nii O. Attoh-Okine.
© 2017 John Wiley & Sons, Inc. Published 2017 by John Wiley & Sons, Inc.

normal, and the fact that distributions can be fully described by their marginal distributions (Accioly and Chiyoshi, 2004).

Linear correlation has been used as an appropriate measure of dependency of variables following a multivariate normal or elliptical distribution of individual variables. Unfortunately, individual infrastructure condition variables may have fat tails, skewness, and nonnormal features. This makes the inference regarding the dependence based on correlation very misleading for non-Gaussian data (Hu, 2006).

The copula approach has been identified as an approach that can be applied to infrastructure modeling. With the copula approach, the modeling of infrastructure data can be divided into various steps (Roch and Alegre, 2006; Zhang and Singh, 2006):

- Modeling the marginal distributions
- Modeling the dependence structure between marginal distributions
- Identification of joint distribution
- Identification of the copula and its parameters
- Comparisons of the performance of different joint distributions

The copula application in infrastructure systems provides the following (Embrechts et al., 2001):

- A comprehensive understanding and interpretation of dependence
- A strong basis for flexible techniques for simulating dependence of random vectors
- Scale-invariant measures of association, better than the traditional linear correlation
- An objective basis for constructing multivariate distributions fitting the observed data
- A new direction in studying the effect of different dependence structures for functions of dependent random variables

Denote z_n as the data from the nth sensor and $z := \{z_n\}_{n=1}^N$ as the heterogeneous data set; the marginals $z := \{z_n\}_{n=1}^N$ are generally nonidentically or heterogeneously distributed. In most of the sensor applications, the problems are modeled as multisensor data fusion, distribution estimator, or Dirichlet detection. In copula analysis, the joint probability density function (pdf) $f(\mathbf{z})$ of the heterogeneous data set is z pdf $z := \{z_n\}_{n=1}^N$ using Sklar's theorem.

Theorem 7.1 *Let F be an N-dimensional cumulative distribution function (cdf) with continuous marginal cdfs $F_1, F_2, \ldots, F_N$. Then there exists a unique copula function C that for all $z_1, z_2, \ldots, z_N$ in $[-\infty, +\infty]$*

$$F(z_1, z_2, \ldots, z_N) = C(F_1(z_1), F_2(z_2), \ldots, F_N(z_N)). \tag{7.1}$$

The joint pdf can now be obtained by taking the N-order derivative of Equation 7.1:

$$F(z_1, z_2, \dots, z_N) = \frac{\partial^N}{\partial_{z_1} \partial_{z_2} \cdots \partial_{z_N}} C(F_1(z_1), F_2(z_2), \dots, F_N(z_N))$$

$$= f_p(z_1, z_2, \dots, z_N) C(F_1(z_1), F_2(z_2), \dots, F_N(z_N)), \tag{7.2}$$

where $f_p(z_1, z_2, \dots, z_N)$ denotes the product of the marginal pdfs $\{f(z_n)\}_{n=1}^{N}$ and $c(\cdot)$ is the copula density that weights the product distribution appropriately to incorporate dependence between the random variables (Ding et al., 2013).

After Sklar's theorem, a joint multivariate distribution of n correlated variables $X_1, X_2, \dots, X_n$ can be described through the copula function C (Klein et al., 2016). The copula function is a multivariate distribution of an n-dimensional random vector on the unit cube, that is, $C := [0, 1]^n \to [0, 1]$, and is strictly invariant under the condition of monotonously increasing transformation of $X_1, X_2, \dots, X_n$:

$$F_{X_1, X_2, \dots, X_n}(x_1, x_2, \dots, x_n) = C[F_{X_1}(x_1), F_{X_2}(x_2), \dots, F_{X_n}(x_n)]$$

$$= C(u_1, u_2, \dots, u_n), \tag{7.3}$$

where $u_1, u_2, \dots, u_n \in [0, 1]$ are uniformly distributed random realizations of the variates defined as $u_1 = F_{X_1}(x_1)$, $u_2 = F_{X_2}(x_2)$, $\dots, u_n = F_{X_n}(x_n)$. The joint multivariate density function is expressed via the copula density c as follows:

$$F_{X_1, X_2, \dots, X_n}(x_1, x_2, \dots, x_n) = \frac{\partial^n C(u_1, u_2, \dots, u_n)}{\partial u_1 \partial u_2 \dots \partial u_n} \prod_{i=1}^{n} f_{X_i}(x_i)$$

$$= c(u_1, u_2, \dots, u_n) \prod_{i=1}^{n} f_{X_i}(x_i). \tag{7.4}$$

Copulas offer the following advantages: (a) copulas can aggregate risk from disparaging sources, even in the case of both objective and subjective risk distributions, (b) copulas present a nonlinear approach to modeling of different dependences, and (c) copulas are based on ranks and therefore is invariant under strictly increasing transforms (Choi, 2014). One major disadvantage of copulas is that sometimes it is very difficult to identify which parametric copula to use first for a given data. Salvadori (2007) presented a detailed analysis of copula in extreme events analysis. The author demonstrated that standard models can be improved using the copula methods. The basic idea of the copula models is that a multivariate distribution can be described by marginals plus a dependence function called a copula (Yu and Voit, 2006) that binds the marginal distributions together. Different families of copulas have been proposed and described extensively by Nelsen (1999). A copula function is defined as a binary function $C : [0, 1]^2 \to [0, 1]$, which satisfies the following conditions:

1) $C(u, 0) = C(0, u) = 0$ for any $u \in [0, 1]$.
2) $C(u, 1) = C(1, u) = 0$ for any $u \in [0, 1]$.
3) For all $0 \leq u_1 \leq u_2 \leq 1$ and $0 \leq v_1 \leq v_2 \leq 1$.

$$C([u_1, v_1] \times [u_2, v_2]) = C(u_2, v_2) - C(u_1, v_2) - C(u_2, v_1) + C(u_1, v_1) \geq 0.$$

Nelsen (1999) presented a more rigorous mathematical definition of a copula. Sklar's theorem states the following:

Theorem 7.2 *Let X and Y be random variables with the joint distribution function H and the marginal distribution functions F and G, respectively. Then there exists a copula C such that*

$$H(x, y) = C(F(x), G(y)) \tag{7.5}$$

for all x, y in $\mathbb{R}$. If F and G are continuous, then C is unique. Otherwise, the copula C is uniquely determined on Ran(F) × Ran(G). Conversely, if C is a copula and F and G are distribution functions, then the function H defined by (H(x, y)) is a joint distribution function with margins F and G.

A copula, C, is a function that joins or couples multiple distribution functions to their one-dimensional marginal distribution functions:

$$H_{X,Y}(x, y) = C(F_X(x), F_Y(y)). \tag{7.6}$$

Let X and Y be a pair of random variables with cdfs of $F_X(x)$ and $F_Y(y)$, respectively. Also, let their joint cdf be $H_{X,Y}(x, y)$. Each pair, (x, y), of real numbers leads to a point $(F_X(x), F_Y(y))$ in the unit square $[0, 1] \times [0, 1]$, and this ordered pair, in turn, corresponds to a number, $H_{X,Y}(x, y)$, in $[0, 1]$. This correspondence is indeed a function, which is known as a copula (Nelsen, 2006).

Nelsen (1999) proved Sklar's theorem. The theorem can be extended to n dimensions:

$$H(x_1, \ldots, x_n) = C(F_1(x_1), \ldots, F_n(x_n)) \tag{7.7}$$
$$H(x_1, \ldots, x_n) = C(u_1, \ldots, u_n). \tag{7.8}$$

For example, in the bivariate case, if $u = F(x)$ and $v = G(y)$, then $C(u, v)$ captures the dependence structure of $H(x, y)$:

$$C(u, v) = H(F^{-1}(u), G^{-1}(v)). \tag{7.9}$$

In this case, each copula is bounded by the Frechet–Hoeffding lower $W(u, v) = \max(u + v - 1, 0)$ and upper $M(u, v) = \min(u, v)$ bounds that represent perfect (negative and positive) dependence, and $\Pi(u, v) = uv$ means perfect independence. This leads to the following: $W(u, v) \leq C(u, v) \leq M(u, v)$. Most copula applications are concerned with bivariate data mainly because relatively few copula families have practical n-dimensional generalizations (Huard et al., 2006). The symbols x and y denote the observation of random variables X and Y; u and v denote their marginal cdfs. Therefore, x and y could be any real numbers, but u and v must be in $[0, 1]$. The density (pdf) of a copula is given by

$$c(u, v) = \frac{\partial C(u, v)}{\partial u \partial v}. \tag{7.10}$$

Equations 7.7 and 7.9 represent a powerful relationship that provides linkages between the marginal distributions (densities and their joint distribution). For independent random variables, Equation 7.10 is equal to one. The copula can allow $(n - 1)$ different dependence structures. It is impractical to do this. The best approach is to analyze the data pair by pair using 2 copulas (Huard et al., 2006). Another important issue is that conditional distributions can be expressed from copulas. Generally, there are two major types of copula: (a) elliptical copulas and (b) Archimedean copulas. Elliptical copulas are related to elliptical distributions and have properties similar to a multivariate normal distribution. Two examples of elliptical distributions are the multivariate Gaussian copula and Student's t copula. Archimedean copulas present several desired properties: they are symmetric and associative, and the dependence measure is slightly simplified (Favre et al., 2004).

7.1.1 Archimedean Copulas

Archimedean copulas are the most popular copulas and have a wide range of applications because of Bacigal and Komornikova (2006) the following: (a) the ease with which they can be constructed, (b) the great families of copulas that belong to this class, (c) the many comprehensive properties of the members of this class, (d) the ease to construct, and (e) the ability that can be applied whether the correlations between variables are negative or positive. In the Archimedean copula, the unit square has a dependence function of the form (Genest and Rivest, 1993)

$$C_\phi(u, v) = \phi^{-1}(\phi(u) + \phi(y)); \text{ for } u, v \in (0, 1]. \tag{7.11}$$

Assume ϕ is a convex, decreasing function with domain $(0,1]$ and range in $[0,\infty]$, that $(0,1] \longrightarrow [0, \infty]$, such that $\phi(1) = 0$. The Archimedean copula is symmetric and associative. Differentiating the generator twice, the copula density function is

$$c_\phi(u, v) = \frac{\partial^2 C\phi(u, v)}{\partial u \partial v} = \frac{-\phi''(C_\phi(u, v))\phi'(u)\phi'(v)}{[\phi'(C_\phi(u, v))]^3} \tag{7.12}$$

Archimedean copulas have closed form expressions for both the copula function and copula density function Wang et al. (2008). One parameter families of the Archimedean class can be summarized as follows (C_ϕ is replaced by C_θ) (Bacigal and Komornikova, 2006). Table 7.1 shows different Archimedean families.

Table 7.1 Archimedean copulas.

Family of copulas	Generator $\phi(t)$	Parameter θ	Bivariate copula $C_\phi(u,v)$
Independence	$-\ln t$		uv
Gumbel	$(-\ln\ t)^\theta$	$\theta \geq 1$	$e^{-[(-\ln\ u)^\theta+(-\ln\ v)^\theta]^{-1/\theta}}$
Clayton	$t^{-\theta}-1$	$\theta > 0$	$(u^{-\theta}+v^{-\theta}-1)^{-\frac{1}{\theta}}$
Frank	$-\ln\left(\dfrac{e^{-\theta t}-1}{e^{-\theta}-1}\right)$	$\theta \in \mathbb{R}$	$\dfrac{1}{\theta}\ln\left(1+\dfrac{(e^{-\theta u}-1)(e^{-\theta v}-1)}{(e^{-\theta}-1)}\right)$

For example, what if one would like to know the probability that two pavement condition variables, cracks and rutting, are in their lowest 10th percentiles? Using an independent bivariate copula, the following expression can be used: $(u,v) = uv = C(0.1,0.1) = 0.01$. With the Archimedean copula, the following properties are very important: Clayton copulas are strong for left tail dependence, Gumbel copulas are useful for highly correlated variables at high values and less correlated values at low levels, and Frank copulas tend to work well when tail dependence is very weak. Therefore, Gumbel and Clayton copulas represent only positive dependence.

7.1.1.1 Concordance Measures

The most common measures used to analyze dependence are the (a) Pearson, (b) Spearman, and (c) Kendall coefficients. The Pearson coefficient appears to be appropriate for demonstrating linear relationships between variables and is not invariant under nonlinear transformation. The Spearman coefficient is similar to the Pearson, but it is calculated using ratings of values rather than the actual values. The Kendall, although similar to the rank-order approach, is distribution-free and does not require the relationship to be linear. Concordance measures of dependence are based on a form of dependence known as concordance (Yan, 2005). The most used concordance measures are Kendall's (τ) and Spearman's (ρ).

The Kendall's τ measure (τ) of pair (x,y), distributed according to H, can be defined as the difference between probabilities of concordance and discordance for two independent pairs (x_1, y_1) and (x_2, y_2), each with the distribution H:

$$\tau(x,y) = \Pr((x_1-x_2)(y_1-y_2) > 0) - \Pr((x_1-x_2)(y_1-y_2) < 0). \quad (7.13)$$

$(x_1-x_2)(y_1-y_2) > 0$ is referred to as concordant, and $(x_1-x_2)(y_1-y_2) < 0$ is referred to as discordant. Kendall's τ can be estimated as follows:

$$\tau(\widehat{x,y}) = \frac{\#\text{Concordant pairs} - \#\text{discordant pairs}}{\#\text{pairs}}. \quad (7.14)$$

Kendall's τ can be expressed through copulas as follows:

$$\tau(x, y) = \tau(C) = 4 \int_0^1 \int_0^1 C(u, v) dC(u, v) - 1. \tag{7.15}$$

For Archimedean copulas, Kendall's τ can be written as

$$\tau_C = 1 + 4 \int_0^1 \frac{\phi(t)}{\phi'(t)} dt. \tag{7.16}$$

Kendall's τ can be considered a measure of the degree of monotonic dependence between x and y. The following are properties of Kendall's τ (Djehiche et al., 2004):

- Insensitive to outliers
- Able to measure the "average dependence" between x and y
- Invariant under strictly increasing transformations

Using a similar presentation to Kendall's τ, Spearman's ρ is

$$\rho_C = 12 \int_0^1 \int_0^1 (C(u, v) - uv) du dv. \tag{7.17}$$

Kendall's τ and Spearman's ρ are called *rank correlations*.

Measures of association related to Archimedean copulas are shown in Table 7.2.

$D_k(x) = \frac{k}{x^k} \int_0^x \frac{t_k}{e^t - 1} dt$ is called the "Debye" function.

The general idea is to choose the best copula from a set of estimated ones. Genest and Favre (2007) presented the steps and methods for estimating the best copula models. Nonparametric estimation and semiparametric estimation methods are used. In the nonparametric estimation, one dimensional function $K(z)$ to its parametric estimate $K_\phi(z)$, the closer $K(z) = K_\phi(z)$. The semiparametric estimation method involves transforming the marginal observations into uniformly distributed vectors, using them in the empirical distribution,

Table 7.2 Kendall and Spearman's values.

Family	Kendall's τ	Spearman's ρ
Independence	0	0
Gumbel	$\dfrac{\theta - 1}{\theta}$	No closed form
Clayton	$\dfrac{\theta}{\theta + 2}$	Complicated form
Frank	$1 - \dfrac{4}{\theta}\{1 - D_1(\theta)\}$	$1 - \dfrac{12}{\theta}\{D_1(\theta) - D_2(\theta)\}$

and estimating copula parameters by maximizing a pseudolikelihood function. The pseudolikelihood function is as follows:

$$L(\theta) = \sum_{i=1}^{n} \log(c_\theta(F_n(x), G_n(y))). \tag{7.18}$$

c_θ is equal to $c_\phi(u, v) = \frac{\partial^2 C\phi(u,v)}{\partial u \partial v} = \frac{-\phi''(C_\phi(u,v))\phi'(u)\phi'(v)}{[\phi'((C_\phi(u,v))]^3}$. The goodness estimation is based on the Akaike information criterion (AIC),

$$AIC = -2(\log-\text{likelihood}) + 2k, \tag{7.19}$$

where k is the number of parameters. The lowest AIC determines the best estimator. The steps in selecting ϕ are as follows (Purcaru and Goegebeur, 2003; Bacigal and Komornikova, 2006; Zhang and Singh, 2006):

- Estimate Kendall's tau (τ) using the nonparametric estimator:

$$\hat{\tau} = \binom{n}{2}^{-1} \sum_{i=2}^{n} \sum_{j=1}^{i-1} \text{sign}[(x_i - x_j)(y_i - y_j)]$$

Determine the copula parameter θ for each copula; use these values to generate ϕ, and obtain the copula from its generating function. The identification of the appropriate copula can be developed based on the following steps (Genest and Rivest, 1993):

- Construct a nonparametric estimate of K. Define a pseudovariable Z_i, having a distribution function $K(z) = \Pr(Z_i \leq z)$, as

$$Z_i = \frac{\text{number of pairs } (x_j, y_j) \text{ in the sample such that}}{n - 1}$$
$$\frac{x_j < x_i \text{ and } y_j < y_i}{n - 1}$$

$$K_n(z) = \frac{1}{n} \sum_{i=1}^{n} \text{if } [Z_i < z, 1, 0]. \tag{7.20}$$

- Construct a parametric estimate of $K_\phi(z)$ using

$$K_\phi(z) = z - \frac{\phi(z)}{\phi'(z)}. \tag{7.21}$$

- Compare $K_n(z)$ and $K_\phi(z)$. Measuring closeness can be done as follows:

$\int_0^1 [K_{\phi n}(z) - K_n(z)]^2 dz$. This can also be achieved by plotting quartile–quartile ($Q - Q$) plots.

The plot of $K_n(z)$ and $K_\phi(z)$ approximates a straight line that passes $(0,1)$, and this shows that the parameter estimation is good enough, and if the Euclidean distance

$$d = \| K_n(z) - K_\phi(z) \|, \tag{7.22}$$

the lower the distance, the better the copula fits.

7.1.2 Multivariate Archimedean Copulas

Nested copulas are another approach used to build multivariate copulas. The bivariate Archimedean copulas can be extended to the d-dimensional case. For every $d \geq 2$ the function C: $[0,1]^d \rightarrow [0, 1]$ is defined as

$$C(u) = \phi^{-1}(\phi(u_1) + \phi(u_2) + \phi(u_3) + \phi(u_4) + \phi(u_5) + \cdots + \phi(u_d)).$$

Various authors, Savu and Trede (2006); Schirmacher and Schirmacher (2008); Aas and Berg (2009), presented multivariate copulas. Aas and Berg (2009) have presented nested Archimedean copulas. These types of copulas join two or more ordinary bivariates or higher dimensional Archimedean copulas with another Archimedean copula (Savu and Trede, 2006). Aas and Berg (2009) presented the following nested Archimedean copulas:

- Fully nested Archimedean copula (FNAC) that allows the specification of the $d - 1$ copula
- Partially nested Archimedean copula (PNAC)
- Hierarchically nested Archimedean copula (HNAC)

The FNAC 4-copula can be represented as follows:

$$C(u) = \phi_3^{-1}[(\phi_3(\phi_2^{-1}[\phi_2(\phi_1^{-1}[\phi_1(u_1) + \phi_1(u_2)] + (\phi_2(u_3) + \phi_3(u_4)].$$

The construction of the PNAC is straightforward; in the case of $d = 4$,

$$C(u) = \phi^{-1}[(\phi(\phi_{12}^{-1}[\phi_{12}(u_1) + \phi_{12}(u_2)]) + \phi(\phi_{34}^{-1}[\phi_{34}(u_3) + \phi_{34}(u_4)]).$$

The basic framework of the HNAC is based on nested multivariate Archimedean copulas. Each copula from a previous level is aggregated until one single copula ends at the top of the hierarchy. The HNACs at specific levels do not have to be bivariate. The basic idea of HNAC is to develop a hierarchy of Archimedean copulas with L different levels. At each level, there are n_l distinct objects. At level $l = 1$, the variables $u_1, \ldots, u_d$ are grouped into n_1 ordinary multivariate Archimedean copulas. These copulas are, in turn, coupled into n_2 copulas at level $l = 2$ (Savu and Trede, 2006).

The HNAC allows the specification of max $d - 1$ copulas, while the remaining unspecified copulas are implicitly obtained from the construction process (Aas and Berg, 2009):

- All inverse generator functions must be monotone.
- Degree of dependence must decrease with level of nesting.
- All bivariate copulas must belong to the same family.

Determination of multivariate dependence measures is not straightforward, and various authors have presented empirical approaches to determine the measures (Joe, 1996; Nelsen, 2006). Unfortunately, it has been shown that the simulation of multivariate copulas is very inefficient for high dimensions (Aas and Berg, 2009).

To address the complexity of simulation of multivariate copulas in higher dimensions, the pair copula construction approach was proposed by Joe (1996). The pair construction allows the specification of $d(d-1)/2$ bivariate copulas of which the $(d-1)$ are unconditional and the rest are conditional (Aas and Berg, 2009). One important property of the pair construction is that the bivariate copulas involved do not have to belong to the same class.

7.2 Pair Copula: Vines

Standard multivariate copulas have the following problems (Krämer and Schepsmeier, 2011):

- They can be inflexible in high dimensions.
- They do not allow dependency structures between pairs of variables.

The vine copula, as explained in detail by Aas et al. (2006), is a flexible graphical model for describing multivariate copulas built up using a cascade of bivariate copulas, where each pair-copula can be chosen independently from the others.

This pair-copula construction is sometimes referred to as vine copulas. It consists of step-by-step factorization of the density function in p product bivariate copulas, and it appears to be more effective than the nested copulas proposed by Aas et al. (2006).

The vine is a graphical model for dependent random variables. A major difference between vines and other graphical models is that vines are conditionally independent, with a conditional correlation coefficient that depends on the value of the root node (Bedford and Cooke, 2001).

Using Sklar's theorem, every multivariate distribution F, with marginals F_1, $F_2, \dots, F_n$ can be expressed as follows:

$$f(x_1, \dots, x_n) = c_{1 \dots n}(F_1(x_1), \dots, F_n(x_n)) \prod_{i=1}^{n} f_i(x_i), \tag{7.23}$$

where $c_{1 \dots n(.)}$ is an n-variate copula. In a bivariate case,

$$f(x_1, x_2) = c_{12}(F_1(x_1), F_2(x_2)).f_1(x_1).f_2(x_2), \tag{7.24}$$

where $c_{12(.)}$ is the appropriate pair-copula density for the pair of transformed variables $F_1(x_1)$ and $F_2(x_2)$. For conditional density, the following equation follows the same pair of copulas:

$$f(x_1 | x_2) = c_{12}(F_1(x_1), F_2(x_2)).f_1(x_1) \tag{7.25}$$

Hobæk Haff and Segers (2015) present the following:

Write $C^{[j]}(u_1, u_2) = \partial C(u_1, u_2)/\partial u_j$ for $j \in \{1, 2\}$. The corresponding conditional distribution function satisfies

$$
\begin{aligned}
F_{1|2}(x_1 \mid x_2) &= \int_{-\infty}^{x_1} f_{1|2}(z \mid x_2) dz = \int_{-\infty}^{x_1} c\{F_1(z), F_2(x_2)\} f_1(z) dz \\
&= \int_0^{F_1(x_1)} c\{u, F_2(x_2)\} du = \frac{\partial}{\partial u_2} C(u_1, u_2)|_{(u_1, u_2) = (F_1(x_1)), F_2(x_2))} \\
&= C^{[2]}\{F_1(x_1), F_2(x_2)\}.
\end{aligned}
\tag{7.26}
$$

Three-dimensional pair-copula decomposition can be expressed as follows:

$$
f(x_1, x_2, x_3) = f(x_1).f(x_2|x_1).f(x_3|x_1, x_2). \tag{7.27}
$$

$$
f(x_2|x_1) = c_{12}(F_1(x_1), F_2(x_2)).f_2(x_2). \tag{7.28}
$$

$$
f(x_3|x_1, x_2) = \frac{f(x_2, x_3|x_1)}{f(x_2|x_1)}. \tag{7.29}
$$

$$
f(x_3|x_1, x_2) = \frac{c_{23|1}(F(x_2|x_1)F(x_3|x_1)f(x_2|x_1))f(x_3|x_1)}{f(x_2|x_1)}. \tag{7.30}
$$

$$
f(x_3|x_1, x_2) = c_{23|1}(F(x_2|x_1), F(x_3|x_1))f(x_3|x_1). \tag{7.31}
$$

$$
f(x_3|x_1, x_2) = c_{23|1}(F(x_2|x_1), F(x_3|x_1)). c_{13}(F_1(x_1), F_3(x_3)).f_3(x_3). \tag{7.32}
$$

A general equation can be developed as follows:

$$
f(x|\mathbf{v}) = cxvj|v_{-j}(F(x|\mathbf{v}_{-j}), F(vj|\mathbf{v}_{-j})).f(x|\mathbf{v}_{-j}) \tag{7.33}
$$

for d dimensional vector v, where $\mathbf{v}_{-j}$ denotes the vector v, excluding the jth component.

The pair-copulas can be used in combination in the following situations:

- Gaussian— when there is no tail dependence
- Student's t—when there is upper and lower tail dependence
- Clayton—when there is only lower tail dependence
- Gumbel—when there is only upper tail dependence

In the high-dimensional distribution, there are tendencies to obtain significant numbers of possible pair-copula constructions. Bedford and Cooke (2001, 2002) proposed the use of vines, which embrace a large number of pair-copula constructions. The authors proposed two types of vines: (a) canonical C-vines and (b) D-vines.

N-Dimensional density $(f(x_1, \ldots, x_n))$ corresponding to C-vines can be represented as follows:

$$\prod_{k=1}^{n} f(x_k) \prod_{j=1}^{n-1} \prod_{i=1}^{n-j} c_{j,j+i|1,\ldots,j-1}(F(x_j|x_1, \ldots, x_{j-1}), F(x_{j+i}|x_i, \ldots, x_{j-1})). \quad (7.34)$$

D-vines can be represented as follows:

$$\prod_{k=1}^{n} f(x_k) \prod_{j=1}^{n-1} \prod_{i=1}^{n-j} c_{i,i+j|i+1,\ldots,i+j-1}$$

$$(F(x_i|x_{i+1}, \ldots, x_{i+j-1}), F(x_{i+j}|x_{i+1}, \ldots, x_{i+j-1})). \quad (7.35)$$

For the d-dimensional C-vine, the pairs at level 1 are 1, i, for $i = 2, \ldots, d$ and for level $l(2 \leq l \leq d)$, the (conditional) pairs are $l|i|1, \ldots, l-1$ for $i = l + 1, \ldots, d$. For the d-dimensional D-vine, the pairs at level 1 are $i, i+1$, for $i = 1, \ldots, d-1$, and for level $l(L(2 \leq l \leq d))$, the (conditional) pairs are $i, i+l|i+1, \ldots, I+l-1$ for $1 = 1, \ldots, d-l$. (Nikoloulopoulos et al., 2012).

Bedford and Cooke (2002) presented the simulation algorithm for a vine and the inference for a pair-copula decomposition. The three major elements involved in the inference are ((a) the selection of a specific factorization, (b) the choice of pair-copula types, and (c) the estimation of the parameters of the chosen pair copulas (Aas, 2006). The construction of the pair-copula has to be evaluated as a conditional distribution of the form $F(x|\mathbf{v})$, where $\mathbf{v}$ denotes a vector of variables. Joe (1996) showed that

$$F(x|v) = \frac{\partial C_{x,v_j}|\mathbf{v}_{-j}(F(x|\mathbf{v}_{-j}), F(v_j|\mathbf{v}_{-j}))}{\partial F(v_j|\mathbf{v}_{-j})}. \quad (7.36)$$

$\partial C_{x,v_j}|\mathbf{v}_{-j}$ is a bivariate copula function. In the case in which x and v are uniform, that is, $f(x) = f(v) = 1$, $F(x) = x$ and $F(v) = v$, then

$$F(x|v) = h(x, v; \Theta) = \frac{\partial C_{x,v}(x, v; \Theta)}{\partial v} \quad (7.37)$$

The vine decomposition of a joint density function with n variables involves $\binom{n}{2}$ pair-copulas. The first tree in the representation has $n-1$ edges, and the second tree has $n-1$ nodes and $n-2$ edges. Therefore, to specify a d-dimensional density, the main steps involve the selection of an appropriate decomposition scheme in the present analysis and specification of pair-copulas.

7.3 Computational Example

One mile of track geometry data was obtained from a US Class I railroad. The track geometry variables in the data set include cross-level, surface (right),

Table 7.3 Correlation matrix based on Kendall's tau.

	Cross-level (C)	Surface (right) (S)	Alignment (right) (A)
Cross-level	1	−0.11	0.071
Surface (right)	−0.11	1	0.139
Alignment (right)	0.071	0.139	1
Gage	−0.071	−0.048	−0.201

alignment (right), and gage measurements. Two thousand observations were used in the analysis. Figure 7.1 shows pairs plots of the track geometry data set with scatterplots above the diagonal and contour plots with standard normal margins below the diagonal. Table 7.3 shows the results of the correlation matrix based on Kendall's tau.

Detailed exploratory data analysis was conducted on the pair of variables gage and alignment (right) that were found to have the highest dependence. Illustrative tools such as Kendall's plot (K-plot), the chi-plot, and the lambda function can be employed for detecting dependence of the pair of variables. Figure 7.2 shows the K-plot, chi-plot, empirical lambda function (black line), and theoretical lambda function for Student's t copula distribution, as well as independence and comonotonicity limits (dashed line). The contour plot in row 4, column 3 of Figure 7.1 as well as the Kendall and chi-plots in Figure 7.1 show that the variables are negatively dependent.

7.3.1 Results

Four-dimensional C-vine and D-vine copula models were applied to the track geometry data. The data set was first transformed into marginally uniform data in the unit interval $[0, 1]$ by applying the empirical distribution functions to the data and scaling the result by $\frac{n}{n+1}$.

The order of the root nodes and the first tree completely determined the structures of the C-vine and D-vine copula models, respectively. The root node of each tree of the C-vine was determined by finding the node with the strongest dependencies with other nodes. This is achieved by finding the node with the maximum row sum of the absolute values in the empirical Kendall's τ matrix. As shown in Table 7.4, alignment (right) was identified as the root node of the first tree. Subsequently, given the first root node and the sequential C-vine identification procedure outlined in Czado et al. (2013), the next root node was identified to be cross-level (as shown in Table 7.5) followed by gage and finally surface (right).

The structure of the D-vine is determined by establishing the order of the first tree. This can be obtained by finding the path that maximizes the pairwise

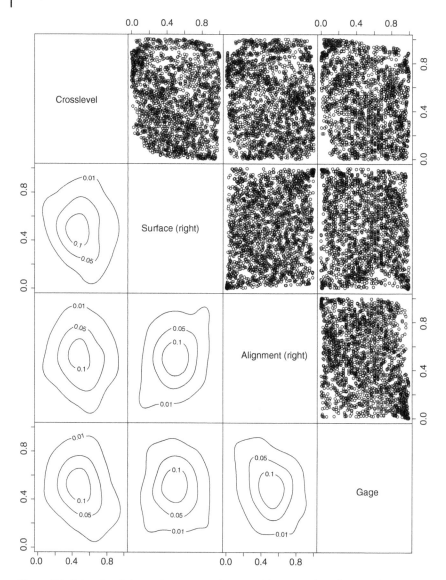

Figure 7.1 Pairs plot of the track geometry data set with scatterplots above and contour plots with standard normal margins below the diagonal

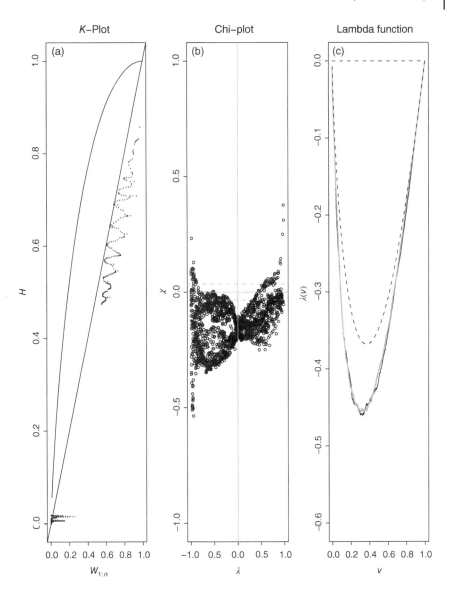

Figure 7.2 (a) *K*-Plot. (b) Chi-plot. (c) Empirical lambda function (black line), theoretical lambda function of a Student's *t* copula (gray line), as well as independence and comonotonicity limits (dashed lines)

Table 7.4 The empirical Kendall's τ matrix and the sum over the absolute entries of each row for the track geometry data set.

	Cross-level (C)	Surface (right) (S)	Alignment (right) (A)	Gage (G)	Sum of absolute τ
Cross-level (C)	1	−0.11	0.071	−0.071	1.252
Surface (right) (S)	−0.11	1	0.139	−0.048	1.296
Alignment (right) (A)	0.071	0.139	1	−0.201	**1.411**
Gage (G)	−0.071	−0.048	−0.201	1	1.321

Table 7.5 The empirical Kendall's τ matrix and the sum over the absolute entries of each row for the derailment data set given alignment (right) (A) as first root.

	A, C	A, S	A, G	Sum of absolute τ
A, C	1	−0.1	−0.23	1.33
A, S	−0.1	1	0.02	**1.12**
A, G	−0.23	0.02	1	1.25

dependences (Kendall's τ) of the variables of interest. In order to obtain the order of the D-vine, a Hamiltonian path is determined.

Furthermore, the sequential tree-by-tree selection via the maximum spanning tree algorithm proposed by Dissmann et al. (2012) for selecting appropriate arbitrary R-vine copulas was performed. The approach establishes the maximum spanning tree for each tree in terms of the absolute empirical pairwise Kendall's τ values while taking into consideration the proximity condition. The resulting structure of the R-vine was found to be similar to that of the D-vine previously selected. This gave an early indication of which of the R-vines might be more appropriate. The pair-copula families considered during the analysis were the Gaussian copula, Student's t copula, Clayton copula, Gumbel copula, Frank copula, and Joe copula. The properties of these copulas are found in Table 7.6.

In order to determine the best-fitting copula to choose, the AIC (Akaike, 1998) was employed, which corrects the log-likelihood of a copula for the number of parameters. The AIC has been found to be a more reliable criterion for bivariate copula selection relative to other criteria such as copula goodness-of-fit tests and the Bayesian information criterion (Dissmann et al., 2012). The independence copula test was included in the selection by performing a preliminary independence test based on Kendall's τ (Genest

Table 7.6 Properties of pair-copula families considered.

Copula	Properties
Normal/Gaussian (N)	Tail-symmetric, no tail dependence
Student's t copula (t)	Tail-symmetric, tail dependence
Clayton (C)	Tail-asymmetric, suitable for modeling lower tail dependence
Gumbel (G)	Tail-asymmetric, upper tail dependence, suitable for highly correlated variables at high values and less correlated values at low levels
Frank (F)	Tail-symmetric, no tail dependence, tends to work well when tail dependence is very weak

Table 7.7 Log-likelihood, number of parameters, AIC, and BIC for C-vine and D-vine copula models using maximum likelihood estimation (MLE) or sequential estimates.

	D-Vine copula model	C-Vine copula model
Log-likelihood (sequential estimation)	461.1109	459.8281
Log-likelihood (MLE)	461.2246	460.3346
Number of parameters	7	7
AIC (sequential estimation)	−908.2217	−905.6561
AIC (MLE)	−908.4491	−906.6693
BIC (sequential estimation)	−869.0154	−866.4498
BIC (MLE)	−869.2428	−867.4629

and Favre, 2007). Pair-copula parameter estimation was performed using the sequential estimation approach proposed by Aas and Berg (2009). The values from sequential estimation were used as starting values for joint maximum likelihood estimation (MLE) (Table 7.7). Figures 7.3 and 7.4 show the C- and D-vine copula models with family and empirical τ values in each tree. They show 3 trees for 4 variables where N is the Normal/Gaussian copula, t is the Student's t copula, F is the Frank copula, and I is the independent copula.

The AIC was used to measure which R-vine structure modeled the data better. The structure with the lower AIC modeled the data better. The D-vine model (Figure 7.4) was found to model the data better than the C-vine model (Figure 7.3). This was confirmed by alternative criteria, such as the Bayesian information criteria and likelihood ratio-based tests, such as the Vuong and Clarke tests.

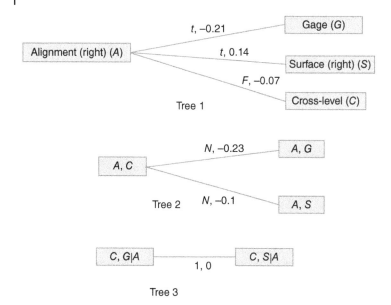

Figure 7.3 Four-dimensional C-vine, where t =Student's t copula, F = Frank copula, N = Normal/Gaussian copula, and I = independent copula with corresponding empirical τ values shown on the links with the copula family

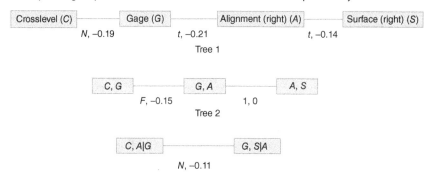

Figure 7.4 Four-dimensional D-vine, where N = Normal/Gaussian copula, t+ Student's t copula, F = Frank copula, and I = independent copula with corresponding empirical tau values shown on the links with the copula family

7.4 Remarks

It is very important to identify the probability distribution of any track (variable) before embarking on further analysis. Also, in some situations where one may not have enough data to make a strong statistical conclusion, a copula can be an appropriate method to generate large data sets that will be statistically equivalent to the "small" data sets.

References

K. Aas and D. Berg. Models for construction of multivariate dependence - a comparison study. *The European Journal of Finance*, **15**(7-8):639–659, 2009. doi: 10.1080/13518470802588767.

K. Aas, C. Czado, A. Frigessi, and H. Bakken. Pair-copula constructions of multiple dependence. *Insurance: Mathematics and Economics*, **44**(2):182–198, 2006. doi: 10.1016/j.insmatheco.2007.02.001.

R. M. S. Accioly and F. Y. Chiyoshi. Modeling dependence with copulas: a useful tool for field development decision process. *Journal of Petroleum Science and Engineering*, **44**(1-2):83–91, 2004. doi: 10.1016/j.petrol.2004.02.007.

H. Akaike. *Information Theory and an Extension of the Maximum Likelihood Principle.* pages 199–213. Springer, New York, 1998. doi: 10.1007/978-1-4612-1694-0_15.

N. O. Attoh-Okine. Pair-copulas in infrastructure multivariate dependence modeling. *Construction and Building Materials*, **49**:903–911, 2013. doi: 10.1016/j.conbuildmat.2013.06.055.

T. Bacigal and M. Komornikova. Fitting Archimedean copulas to bivariate geodetic data. In *Compstat Satellite Meetings*, pages 1–8, 2006. http://www.stat.unipg.it/iasc/Proceedings/2006/COMPSTAT/CD/251.pdf.

T. Bedford and R. M. Cooke. Probability density decomposition for conditionally dependent random variables modeled by vines. *Annals of Mathematics and Artificial Intelligence*, **32**(1-4):245–268, 2001. doi: 10.1023/A:1016725902970.

T. Bedford and R. M. Cooke. Vines–a new graphical model for dependent random variables. *The Annals of Statistics*, **30**(4):1031–1068, 2002. http://projecteuclid.org/euclid.aos/1031689016.

J. Choi. *Qualitative analysis for dynamic behavior of railway ballasted track.* PhD thesis, TU Berlin, 2014.

C. Czado, E. C. Brechmann, and L. Gruber. *Selection of Vine Copulas.* pages 17–37. Springer, Berlin Heidelberg, 2013. doi: 10.1007/978-3-642-35407-6_2.

G. Ding, L. Wang, and Q. Wu. Big data analytics in future internet of things. *arXiv.org*, cs.DC(61172062):1–6, 2013. http://arxiv.org/abs/1311.4112v1$npapers3://publication/uuid/1ACE7CC5-49C8-459E-8BD6-6B88A26D8E0B.

J. Dissmann, E. C. Brechmann, C. Czado, and D. Kurowicka. Selecting and estimating regular vine copulae and application to financial returns, 2012. http://arxiv.org/abs/1202.2002.

B. Djehiche, S. Liv, and H. Hult. An introduction to copulas with applications, 2004.

P. Embrechts, F. Lindskog, and A. McNeil. Modelling dependence with copulas and applications to risk management. Technical report, 2001. http://citeseerx.ist.psu.edu/viewdoc/summary?doi=10.1.1.23.5130.

A.-C. Favre, S. El Adlouni, L. Perreault, N. Thiémonge, and B. Bobée. Multivariate hydrological frequency analysis using copulas. *Water Resources Research*, **40**(1), 2004. doi: 10.1029/2003WR002456.

C. Genest and A.-C. Favre. Everything you always wanted to know about copula modeling but were afraid to ask. *Journal of Hydrologic Engineering*, **12**(4):347–368, 2007. doi: 10.1061/(ASCE)1084-0699(2007)12:4(347).

C. Genest and L.-P. Rivest. Statistical inference procedures for bivariate Archimedean copulas. *Journal of the American Statistical Association*, **88**(423), 1993. doi: 10.1080/01621459.1993.10476372.

I. Hobæk Haff and J. Segers. Nonparametric estimation of pair-copula constructions with the empirical pair-copula. *Computational Statistics & Data Analysis*, **84**:1–13, 2015. http://www.sciencedirect.com/science/article/pii/S0167947314003120.

L. Hu. Dependence patterns across financial markets: a mixed copula approach. *Applied Financial Economics*, 717–729, 2006. http://citeseerx.ist.psu.edu/viewdoc/summary?doi=10.1.1.198.6725.

D. Huard, G. Évin, and A.-C. Favre. Bayesian copula selection. *Computational Statistics & Data Analysis*, **51**(2):809–822, 2006. doi: 10.1016/j.csda.2005.08.010.

H. Joe. Families of m-variate distributions with given margins and (m-1)/2 bivariate dependence parameters. In *Lecture Notes–Monograph Series*, volume **28**, pages 120–141. Institute of Mathematical Statistics, 1996. ISBN: 0-940600-40-4. http://projecteuclid.org/euclid.lnms/1215452614.

B. Klein, D. Meissner, H.-U. Kobialka, and P. Reggiani. Predictive uncertainty estimation of hydrological multi-model ensembles using pair-copula construction. *Water*, **8**(4):125, 2016. ISSN: 2073-4441.

N. Krämer and U. Schepsmeier. Introduction to vine copulas, 2011.

R. B. Nelsen. *An Introduction to Copulas*. Springer Science & Business Media, 1999. ISBN: 0387986235. https://books.google.com/books/about/An_Introduction_to_Copulas.html?id=5Q7ooTrVe9sC&pgis=1.

R. B. Nelsen. *An Introduction to Copulas*, volume **53**. Springer, New York, 2nd edition, 2006. ISBN: 9788578110796.

A. K. Nikoloulopoulos, H. Joe, and H. Li. Vine copulas with asymmetric tail dependence and applications to financial return data. *Computational Statistics & Data Analysis*, **56**(11):3659–3673, 2012. doi: 10.1016/j.csda.2010.07.016.

O. Purcaru and Y. Goegebeur. Modelling dependence through copulas, 2003.

O. Roch and A. Alegre. Testing the bivariate distribution of daily equity returns using copulas. An application to the Spanish stock market. *Computational Statistics & Data Analysis*, **51**(2):1312–1329, 2006. doi: 10.1016/j.csda.2005.11.007.

G. Salvadori. *Extremes in Nature: An Approach Using Copulas*. Springer Science & Business Media, 2007. ISBN: 1402044143. https://books.google.com/books?id=xpimMh2JSvwC&pgis=1.

C. Savu and M. Trede. Hierarchical Archimedean copulas. In *International Conference on High Frequency Finance*, 2006. https://www.google.com/_/chrome/newtab?espv=2&ie=UTF-8.

D. Schirmacher and E. Schirmacher. Multivariate dependence modeling using pair-copulas. In *2008 ERM Symposium*, pages 1–52, 2008. http://www.ermsymposium.org/2008/pdf/papers/Schirmacher.pdf http://www.ermsymposium.org/2008/index.php.

B. Vaz de Melo Mendes and R. Martins de Souza. Measuring financial risks with copulas. *International Review of Financial Analysis*, **13**(1):27–45, 2004. doi: 10.1016/j.irfa.2004.01.007.

M. Wang, K. Rennolls, and S. Tang. Bivariate distribution modeling of tree diameters and heights: dependency modeling using copulas. *Forest Science*, **54**(3):284–293, 2008.

J. Yan. Multivariate modeling with copulas and engineering applications. In *Springer Handbook of Engineering Statistics*, pages 973–990. 2005. ISBN: 978-1-84628-288-1.

L. Yu and E. O. Voit. Construction of bivariate S-distributions with copulas. *Computational Statistics & Data Analysis*, **51**(3):1822–1839, 2006. doi: 10.1016/j.csda.2005.11.021.

L. Zhang and V. P. Singh. Bivariate flood frequency analysis using the copula method. *Journal of Hydrologic Engineering*, **11**(2):150–164, 2006. doi: 10.1061/(ASCE)1084-0699(2006)11:2(150).

8

Topological Data Analysis

8.1 Introduction

Topological data analysis (TDA) is a data-driven approach that involves the study of high-dimensional data without any assumptions or feature selections (Zomorodian, 2011). The major and fundamental idea is that the topological methods act as a geometric approach to patterns and shapes within the data. The shape in the data has meaning; therefore, extracting shapes (patterns) of the data may provide qualitative and quantitative summaries of the data. Most methods of obtaining information from data involve the following steps: (a) one creates a model, (b) one experiments to obtain the data, and (c) the data are inspected to check if they fit the expected model. This works very for areas where there are established theories. In the current age of "big data," data are collected without a particular hypothesis to test. For many complex data sets, especially rail track monitoring, the number of possible hypotheses is very large, and the talk of generating useful ones becomes extremely difficult. Topology therefore can offer a different strategy for big data.

The high-dimensional data are prone to usual amount of noise and errors. Also, the interpretation of the data is tied to the scale at which they are considered. Furthermore, the data can be streamed in high dimensions, which can cause the "curse of dimensionality" problems. Therefore, there is a need to extract robust, qualitative information and gain insight into the processes that generated the data in the first place (Nanda and Sazdanović, 2014).

8.2 Basic Ideas

8.2.1 Topology

Topology is the study of basic properties of a space or object, such as connectedness of the presence or "holes" in the space, that are preserved under

Big Data and Differential Privacy: Analysis Strategies for Railway Track Engineering, First Edition. Nii O. Attoh-Okine.
© 2017 John Wiley & Sons, Inc. Published 2017 by John Wiley & Sons, Inc.

continuous deformation. There are three key ideas to topology that make extracting patterns via shape viable (Lum et al., 2013):

- Topology studies shapes in a coordinate-free way. The topological construction does not depend on the coordinate system chosen but only on the distance function that specifies the shape. This allows topology to compare data derived from different platforms (including different coordinate systems).
- Topology studies the properties of shapes that are invariant under "small" deformation. This means that a circle, an ellipse, and a boundary hexagon are the same under topological invariance. Therefore, a coffee mug is topologically equivalent to a doughnut.
- Topology provides a framework for compressed representation of shapes.

Lum et al. (2013) showed that TDA is more effective in detecting large and small patterns in data compared with traditional principal component analysis (PCA) or cluster analysis.

8.2.2 Homology

Homology is the mathematical prescription that calculates the algebraic properties of objects called chain complexes (these will be discussed in the next few sections). Chain complexes are made of simplices (Fedus et al., 2015).

A simplicial k-chain is a sum of k-simplices (σ_k):

$$c_k = \sum_i \alpha_i \sigma_k^i, \quad \alpha_i \in \mathbb{F}, \tag{8.1}$$

where $\mathbb{F}$ is some field. Each k-simplex can be thought of as a k-dimensional polytope. Thus, a 2-simplex represents a triangle, a 3-simplex represents a tetrahedron, and so on. The boundary operator $\partial_k :\to C_{k-1}$ is a linear homomorphism defined to act on $\sigma_k = [v_0, v_1, \dots, v_k]$:

$$\partial_k \sigma_k = \sum_i (-1)^i [v_0, v_1, \dots, \hat{v}_i, \dots, v_k] \in C_{k-1}, \tag{8.2}$$

where "$\hat{v}_i$" means that this element is removed from the simplex. This definition forces a flow of information in the various chain groups to be allowed:

$$\cdots \to C_{k+1} \xrightarrow{\partial_{k+1}} C_k \xrightarrow{\partial_k} C_{k-1} \to \cdots \tag{8.3}$$

Various subgroups of this map can be defined. In particular, the cycle group $Z_k \equiv \ker \partial_k$, and the boundary group $B_k \equiv \operatorname{im} \partial_{k+1}$. The homology group can be defined as the quotient group:

$$H_k \equiv Z_k / B_k = \ker \partial_k / \operatorname{im} \partial_{k+1}. \tag{8.4}$$

Simplices are the simplest polytopes in a given dimension, as described below (Xia et al., 2015). Let $v_0, v_1, \dots, v_p$ be $p + 1$ affinely independent points in a linear space. A p-simplex σ_p is the convex hull of those $p + 1$ vertices, denoted as σ_p = convex $< v_0, v_1, \ldots, v_p$ or shortened as $\sigma_p =< v_0, v_1, \ldots, v_p$. A formal definition can be given as

$$\sigma_p = \left\{ v \mid v = \sum_{i=0}^{p} \lambda_i v_i, \ \sum_{i=0}^{p} \lambda_i \leq 1, \ \forall i \right\}. \tag{8.5}$$

The most commonly used simplices in $\mathbb{R}^3$ are 0-simplex (vertex), 1-simplex (edges), 2-simplex (triangle), and 3-simplex (tetrahedron), as illustrated in Figure 8.1.

Homology counts components, holes, voids, and so on.

- 0 – simplex point Δ^0
- 1 – simplex line segment Δ^1
- 2 – simplex triangle Δ^2
- 3 – simplex tetrahedron Δ^3

8.2.2.1 Simplicial Complex
A simplicial complex is a finite collection of a set of simplices, as shown in Figure 8.2.

Simplicial homology falls under the following headings:

- Simplicial homology
  - Simplex, simplicial complex
  - Chain group, cycle group, boundary group
  - Homology group, homology class, Betti number

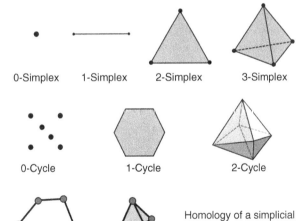

Figure 8.1 Illustration of simplex

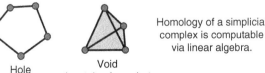

Figure 8.2 Simplicial complex

Hole

Void
(contains faces but empty interiro)

Homology of a simplicial complex is computable via linear algebra.

The boundary $\partial\sigma$ of an oriented k-simplex σ is a $(k-1)$ chain, defined as the following weighted sum of the facets of σ:

$$\partial[x_0, x_1, \dots, x_k] := \sum_{i=0}^{d} (-1)^i [x_0, x_1, \dots, \widehat{x}_i, \dots, x_k], \qquad (8.6)$$

where $[x_0, x_1, \dots, \widehat{x}_i, \dots, x_k]$ indicates the facet opposite vertex x_i. For example, we have

$$\begin{aligned}
\partial[w, x, y, z] &:= [x, y, z] - [w, y, z] + [w, x, z] - [w, x, y] \\
\partial[x, y, z] &:= [y, z] - [x, z] + [x, y] \\
\partial[x, y] &:= [y] - [x] \\
\partial[x] &:= []
\end{aligned} \qquad (8.7)$$

8.2.2.2 Cycles, Boundaries, and Homology

We now describe two important subgroups of the chain groups $C_k(X)$. A k-cycle is a k-chain α such that $\partial_k \alpha = 0$. A k-boundary is a k-chain α such that $\alpha = \partial_{k+1}\beta$ for some $(k+1)$-chain β.

The k-cycles and k-boundaries define subgroups of $C_k(X)$ called the kth cycle group $Z_k(X)$ and the kth boundary group $B_k(X)$. These subgroups can also be defined in terms of the boundary maps as follows:

$$Z_k(X) := \ker \partial_k \quad \text{and} \quad B_k(X) := \operatorname{im} \partial_{k+1}. \qquad (8.8)$$

In linear algebraic terms, $B_k(X)$ is the row space of the matrix ∂_{k+1}, and $Z_k(X)$ is the right null space (the orthogonal complement of the column space) of ∂_k (Table 8.1). All cycle and boundary groups are free abelian groups, meaning they have the form $\mathbb{Z}^c$ for some integer c. $B_k(X) \trianglelefteq Z_k(X) \trianglelefteq C_k(X)$, where $\trianglelefteq$ denotes a "normal subgroup." (If we use real coefficients instead of integers to define chains, then $B_k(X)$ and $Z_k(X)$ are nested linear subspaces of the real vector space $C_k(X)$.)

We now define an equivalence relation over Z_k. Two k-cycles α and β are homologous if the k-cycle $\alpha - \beta$ is a k-boundary. The equivalence class of

Table 8.1 Equivalent definitions of the cycle and boundary groups.

∂_k	Linear map $C_k \to C_{k-1}$	$n_{k-1} \times n_k$ integer matrix
$B_k(X)$	$\operatorname{im} \partial_{k+1}$	Row space of ∂_{k+1}
$Z_k(X)$	$\ker \partial_k$	Right null space of ∂_k

a k-cycle α is the homology class $[\alpha]$. Addition of homology classes is well defined; for any k-cycles α and β, we have $[\alpha + \beta] = [\alpha] + [\beta]$. Thus, the set of homology classes of k-cycles forms a well-defined group under addition called the kth homology group $H_k(X)$. This group can also be defined as the quotient group of k-cycles mod k-boundaries:

$$H_k(X) := \frac{Z_k(X)}{B_k(X)}. \tag{8.9}$$

The topological features detected by simplicial homology correspond to n-dimensional holes. The number of holes is known as the Betti number. Betti numbers are integers that count how many generators of specific dimension exist at a specific filtration. Figure 8.3 presents examples of Betti numbers.

8.2.3 Persistent Homology

8.2.3.1 Filtration

The aim of persistent homology is to measure the lifetime of certain topological properties of a simplicial complex when simplices are added or removed from it (Horak et al., 2009). The filtration process creates a family of similar copies of the object at different spatial resolutions. Figure 8.4 shows the filtration process.

8.2.4 Persistence Visualizations

8.2.4.1 Persistence Diagrams

- *Persistence Barcode.*
- *Persistence Visualizations.* The persistent homology can be presented in a graph format using a persistence diagram. In the graphical representation (Figure 8.5), each persistent homology corresponds to x and y coordinates that are its birth and death times, respectively. The persistence diagram includes all the points corresponding to persistent homology classes. It can be used to develop a persistence barcode. The difference between birth and death time is called persistence, which quantifies the significance of a topological attribute. Furthermore, the persistence barcode is robust under noise (Freedman and Chen, 2009).

Figure 8.3 Betti numbers

	Point	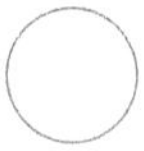 Circle	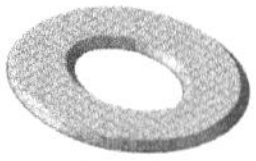 Torus
β_0	1	1	1
β_1	0	1	2
β_2	0	0	1

• Examines topological invariants, for example, homology of the space

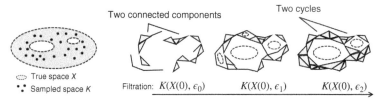

Watch the evolution of simplicial complex *K* increasing the radius, or threshold (ϵ)

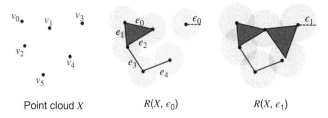

Figure 8.4 Filtration

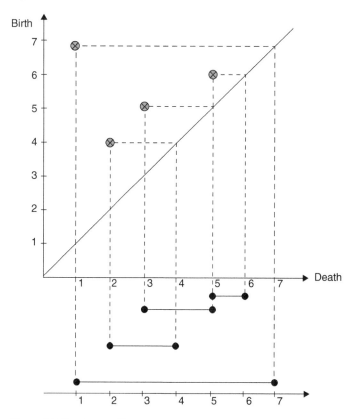

Figure 8.5 Persistence diagram

8.3 A Simple Railway Track Engineering Application

TDA has unlimited applications in:

- Axle box acceleration analysis
- Other signal processing applications

The application of TDA involves using time-delay embedding for signals, and it is used to explore topological properties of one-dimensional time-varying signals. The general representation is as follows:

Given an example (axle box acceleration output) and an embedding dimension n, that is, a scalar time transformed into an n-dimensional space, as shown in Figure 8.6.

8.3.1 Embedding Method

a) Plot $x(t)$ versus $x(t - \tau)$, $x(t - 2\tau)$, $x(t - 3\tau)$.
b) $x(t)$ – any signal (axle box acceleration output, other observable properties).
c) The embedding dimension is the number of delays.
d) The choice of τ and of the dimension is critical.

TDA involves the following steps to convert the signals into a point cloud using time-delay embedding (Krim et al., 2016):

a) Clean the resulting cloud.
b) Compute a topological invariant of the resulting point cloud and use it for classification.

Figure 8.6 Schematic representation of TDA

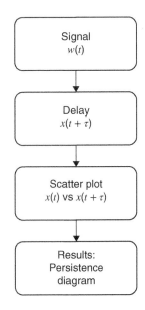

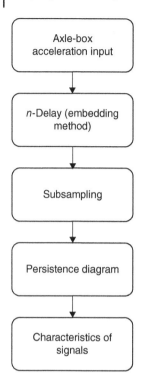

Figure 8.7 Application of TDA

8.4 Remarks

TDA appears to have applications in axle box acceleration analysis. The axle box acceleration input can be transformed using the embedding method, and the resulting graph is analyzed using TDA. Also, the analysis can be used to create a database of different axle box acceleration inputs that can be used for automated analysis (Figure 8.7).

References

W. Fedus, M. Gartner, A. Georges, D. A. Meyer, and D. Rideout. Persistent homology for mobile phone data analysis, 2015.

D. Freedman and C. Chen. Algebraic topology for computer vision, 2009. http://www.hpl.hp.com/techreports/2009/HPL-2009-375.pdf.

D. Horak, S. Maletic, and M. Rajkovic. Persistent homology of complex networks. *Journal of Statistical Mechanics: Theory and Experiment*, **2009**(03):P03034, 2009. doi: 10.1088/1742-5468/2009/03/P03034.

H. Krim, T. Gentimis, and H. Chintakunta. Discovering the whole by the coarse: a topological paradigm for data analysis. *IEEE Signal Processing Magazine*, **33**(2):95–104, 2016. doi: 10.1109/MSP.2015.2510703.

P. Y. Lum, G. Singh, A. Lehman, T. Ishkanov, M. Vejdemo-Johansson, M. Alagappan, J. Carlsson, G. Carlsson, D. N. Reshef, L. Euler, G. Carlsson, Y. Yao, M. Nicolau, A. J. Levine, G. Carlsson, G. Reeb, T. C. Putti, A. E. Teschendorff, A. Miremadi, S. E. Pinder, I. O. Ellis, C. Caldas, T. Sorlie, C. M. Perou, T. Sorlie, D. Venet, J. E. Dumont, V. Detours, L. J. van 't Veer, Y. Wang, L. J. van 't Veer, M. J. van de Vijver, and H. Abdi. Extracting insights from the shape of complex data using topology. *Scientific Reports*, **3**:1518–24, 2013. doi: 10.1038/srep01236.

V. Nanda and R. Sazdanović. Simplicial models and topological inference in biological systems. In N. Jonoska and M. Saito, editors, *Discrete and Topological Models in Molecular Biology, Part of the series Natural Computing Series*, pages 109–141. Springer, Berlin Heidelberg, 2014. ISBN: 978-3-642-40192-3.

L. M. Seversky, S. Davis, and M. Berger. On time-series topological data analysis: new data and opportunities. In *Proceedings of the IEEE Conference on Computer Vision and Pattern Recognition Workshops*, pages 59–67. IEEE, 2016.

K. Xia, X. Feng, Y. Tong, and G. W. Wei. Persistent homology for the quantitative prediction of fullerene stability. *Journal of Computational Chemistry*, **36**(6):408–422, 2015. doi: 10.1002/jcc.23816.

A. Zomorodian. Topological data analysis. In *Proceedings of Symposia in Applied Mathematics*, pages 1–39. American Mathematical Society, 2011. http://citeseerx.ist.psu.edu/viewdoc/summary?doi=10.1.1.261.1298, http://citeseerx.ist.psu.edu/viewdoc/download;jsessionid =E95F73A44B7A00A215292A47FD2B2328?doi=10.1.1.261.1298&rep=rep1 &type=pdf.

9

Bayesian Analysis

9.1 Introduction

In statistical decision-making, there are two main approaches: the frequentist and the Bayesian approaches (Suess and Trumbo, 2010). The frequentist approach assumes that statistical inference can be performed based on experiments that are repeated many times. On the other hand, Bayesian inference is based on personal beliefs about the probability and on observations made from a single instance of an experiment. These two approaches have different methods and their interpretations of the results are therefore different. Although frequentist methods are more settled in scientific research, Bayesian methods have gained importance because they have been shown to provide better solutions in some applications, especially improvements in computation that make them more suitable to practical implementations. Bayesian inference is conceived under the assumption that all the unknown parameters are random variables and their probability distributions denote the beliefs about their values. Markov chain Monte Carlo (MCMC) is one of the computational methods that has been applied widely in Bayesian inference.

In order to estimate the values of the parameter vector θ in Bayesian inference, two elements are required. The first element contains all our beliefs about the parameters before the data is observed; this is called the prior probability distribution. The second element expresses the probability of the data X given the parameters and is called the likelihood. By using Bayes theorem, we can put the prior and the likelihood together, so the probability distribution of the parameters after the data is observed can be obtained; this is called the posterior distribution. In many practical problems, the resultant function computed is intractable by analytical methods, so numerical methods are used instead (Robert, 2014). MCMC is a Bayesian numerical method that simulates via Monte Carlo simulation a Markov chain in such a way that the steady-state distribution of the chain is the posterior distribution of the parameters. In the academic literature, there exist two main approaches to performing

Big Data and Differential Privacy: Analysis Strategies for Railway Track Engineering, First Edition. Nii O. Attoh-Okine.
© 2017 John Wiley & Sons, Inc. Published 2017 by John Wiley & Sons, Inc.

MCMC (Robert, 2014): the Metropolis–Hastings (M–H) algorithm and Gibbs sampling.

9.1.1 Prior and Posterior Distributions

Bayesian inference relies on the assumption that all the parameters (θ) in a specific model are random variables and the data set is known. In order to determine the probability distribution of the model parameters (θ) and, therefore, their estimators, Bayesian inference is supported by Bayes' rule that states the following:

$$\text{posterior} \propto \text{prior} \times \text{likelihood.} \tag{9.1}$$

The joint probability of θ and the data (X) is

$$f(X, \theta) = f(\theta)f(X \mid \theta). \tag{9.2}$$

So the conditional probability of θ given X is

$$f(\theta \mid X) = \frac{f(\theta)f(x \mid \theta)}{\int f(\theta)f(X \mid \theta)d\theta}, \tag{9.3}$$

where the posterior distribution, broadly speaking, is the probability of the parameters after having seen the data $f(\theta \mid X)$. The likelihood is the probability of the data given the parameters $f(x \mid \theta)$. Finally, the prior expresses the beliefs about the parameters before having seen the data $f(\theta)$, where θ is the vector parameter. The choice of the prior is very important because it has an impact on the desired distribution. There are three scenarios that suggest the choice of the prior distribution; each of them has their own consequences, as will be explained below.

The first scenario is known as conjugate priors. The idea in this case is to select a probability distribution that belongs to a family of distributions in such a way that the posterior distribution also belongs to that family. The conjugate prior can be represented as follows:

Let $f(X \mid \theta)$ be the likelihood function that belongs to the probability distribution family A. It can be stated the family A is conjugate with respect to $f(X \mid \theta)$ if the posterior distribution $f(\theta \mid X)$ also belongs to A, regardless of which distribution of θ is in A. One of the issues presented by using non-informative priors is that sometimes the prior distribution is improper, so it could be possible that the posterior distribution would be improper as well. Conjugate priors overcome this issue, because we can select an appropriate distribution such that the posterior probability can be determined. If we know the conjugate priors for a parameter estimation problem, then it is not necessary to perform MCMC routines, because we have an analytical form for the posterior distribution. Table 9.1 presents a summary of some common cases of conjugate priors.

However, identifying the family of distributions is not always an easy task, and we need to rely on other strategies to determine the prior distribution, as shown below:

Table 9.1 Conjugate prior distributions.

Prior	Likelihood	Posterior
Normal	Normal	Normal
Normal	Gamma	Gamma
Bernoulli	Beta	Beta
Poisson	Gamma	Gamma
Binomial	Beta	Beta

When there is no knowledge about the parameters, the literature recommends assuming a weak or uniform prior distribution. This case is known as non-informative priors and, since it apparently tries to represent a conservative position about the parameters, there are mathematical implications (Gelman et al., 2008).

Say we have a prediction model and we want to estimate the parameter values for the normal distribution: $\theta \sim N(\mu, \sigma^2)$ where the parameters of the model are the mean μ and the variance σ^2. If we do not have knowledge about the parameter σ^2, we may want variance to be as spread as possible, so a uniform distribution would be appropriate. The idea is that the prior distribution of the variance will have the minimum impact on the posterior distribution. The problem arises due to the fact that the area under the curve does not sum up to 1. In this case, we say that the prior distribution is improper. Although in the literature it is common to use improper functions, it is important to make sure that the posterior distribution will not be improper as well (Sorensen and Gianola, 2002).

At the other extreme, there are situations in which there is "high" knowledge about the parameter (θ). In this case, the prior represents the beliefs of the community rather than individual ones. This information usually comes from experts' opinions through interviews, surveys, and so on. The information can also come from historical data as a result of computational analysis, past results from engineering tests, and others. Since there is a strong knowledge about the parameter's behavior, the prior is not dominated by the likelihood, so the impact on the posterior distribution is very high. In the literature, these priors are known as informative priors (Sorensen and Gianola, 2002). Because the prior has much influence over the posterior, the assessment of this distribution cannot be the most appropriate due to biases from the experts' opinions. In order to overcome this issue, it is recommended that more data be collected so that the probability assessments would become more objective.

9.2 Markov Chain Monte Carlo (MCMC)

9.2.1 Gibbs Sampling

Let us say that the parameter vector θ can be partitioned into N subvectors:

$$\theta = \begin{bmatrix} \theta_1 \\ \theta_2 \\ \vdots \\ \theta_N \end{bmatrix}.$$

Let $P(\theta_j \mid X, \theta_k, k \neq j)$ be the conditional distribution of θ_j, given the data (X) and the remaining parameter subvectors. If we can sample from each conditional distribution, then Gibbs sampling would be appropriate to implement.

The Gibbs sampling algorithm simulates a Markov chain $x^{(1)}, x^{(2)}, \dots, x^{(k)}$ that converges to $f(x)$. This is obtained by sampling successively the full conditional component distributions $P(\theta_j \mid X, \theta_k, k \neq j)$.

The parameter vector is $\theta = [\alpha, \beta]$; the Gibbs sampling generates the Markov chain

$$\left(\alpha^{(1)}, \beta^{(1)}\right), \left(\alpha^{(2)}, \beta^{(2)}\right), \cdots, \left(\alpha^{(k)}, \beta^{(k)}\right)$$

that converges to $f(\alpha, \beta)$, by successively sampling

$$\alpha^{(1)} \text{ from } f(\alpha \mid X, \beta^{(0)})$$
$$\beta^{(1)} \text{ from } f(\beta \mid X, \alpha^{(1)})$$
$$\vdots$$
$$\alpha^{(k)} \text{ from } f(\alpha \mid X, \beta^{(k-1)})$$
$$\beta^{(k)} \text{ from } f(\beta \mid X, \alpha^{(k)})$$

To start, it is necessary to specify an initial value for $\beta^{(0)}$. The first M iterations depend on the starting value and are discarded from the final solution; this is also known as the burn-in period.

9.2.2 Metropolis–Hastings

The M–H algorithm is the general case of MCMC methods. In general, the M–H algorithm simulates samples from a proposal distribution for each parameter of interest. The proposed candidate is accepted or rejected using the M–H ratio (α) as a criterion.

9.3 Approximate Bayesian Computation

In Bayesian statistics, one needs to evaluate the likelihood function in order to obtain the posterior. Unfortunately, in some cases the likelihood function

---

Algorithm 4 Metropolis–Hastings Algorithm

Select a proposal distribution $q(x_2 \mid x_1)$
Initialize x_1
for $k = 1 \rightarrow$ *iterations* **do**
  Draw $y \sim q(y \mid x_i)$
  $\alpha(x_i, y) \leftarrow \min\left(\frac{p(y)q(y,x_i)}{p(x_i)q(x_i,y)}, 1\right)$
  Draw $U \sim \alpha(x_i, y)$
  if $U < \alpha(x_i, y)$ **then**
    $x_{i+1} \leftarrow y$
  else
    $x_{i+1} \leftarrow x_i$
  end if
end for
return Last N samples

---

is either mathematically difficult to deal with, does not exist, or the computational cost to evaluate it is expensive. In such situations, approximate Bayesian computation (ABC), otherwise known as likelihood-free computation, is an approximate method to use (Beaumont et al., 2002).

ABC has its roots in the rejection algorithm, MCMC, and sequential Monte Carlo, among others (Buzbas and Rosenberg, 2015). This section of the book focuses on the rejection algorithm approach.

Figures 9.1–9.3 shows the comprehensive steps of ABC. Figures 9.4–9.8 show the ABC steps in diagrammatic presentations of the rejection algorithm.

9.3.1 ABC – Rejection algorithm

The rejection algorithm involves the generation of samples from a probability distribution. The basic idea is the algorithm simulating a large number of data sets under hypothetical scenarios. These parameters of the hypothetical scenarios are sampled from a probability distribution. The data generated from the simulation are then reduced to summary statistics, and sampled parameters are either accepted or rejected based on prior selected distance, between simulated and observed data.

Let us assume we have a prior density and some track geometry data g:

1) Choose a set of summary statistics S for observed data $S(q)$.
2) For $k = 1, \ldots, n$.
   a) Simulate θ^* from the prior.
   b) Use θ^* to simulate some artificial data, q^*, and calculate $S(q^*)$.
   c) If $|S(q^*) - S(q)| \leq \epsilon$ accept θ^*.
   d) Else, reject θ^* and repeat (a)-(d).

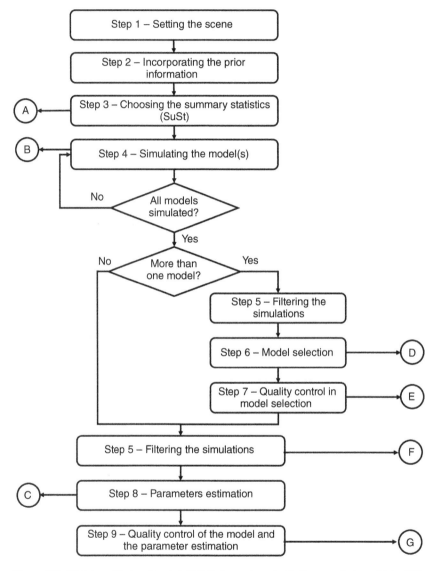

Figure 9.1 ABC steps (Bertorelle et al. (2010). Reproduced with the permission of John Wiley and Sons)

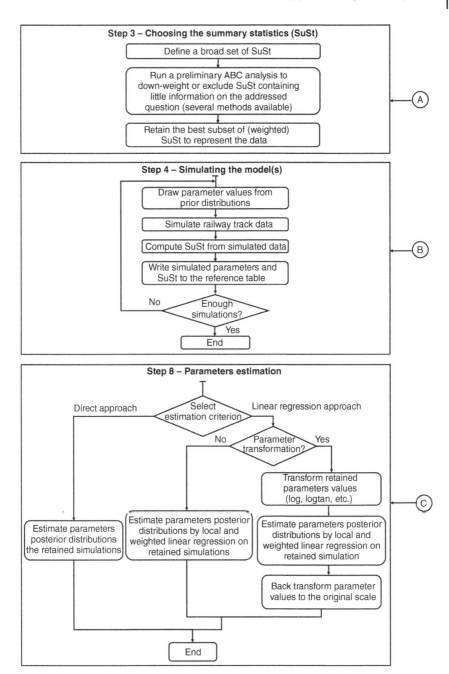

Figure 9.2 ABC steps cont'd (Adapted from: Bertorelle et al. (2010))

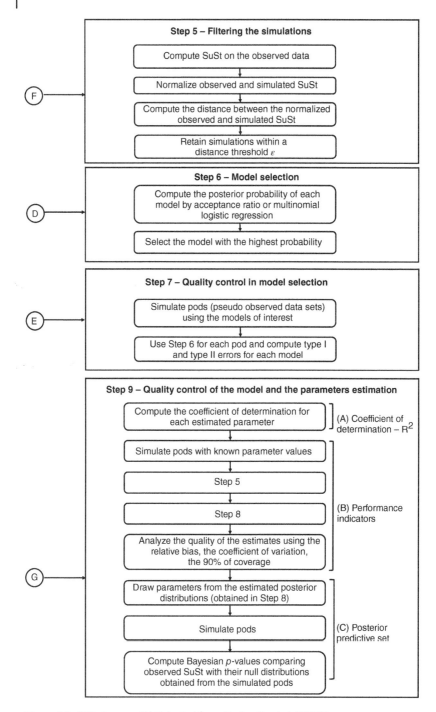

Figure 9.3 ABC steps cont'd (Adapted from: Bertorelle et al. (2010))

Figure 9.4 ABC step 1

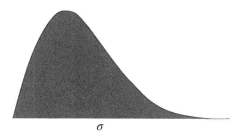

Figure 9.5 ABC step 2

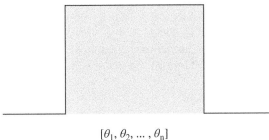

Figure 9.6 ABC step 3

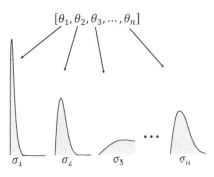

Figure 9.7 ABC step 4

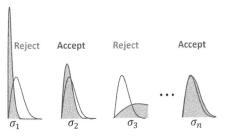

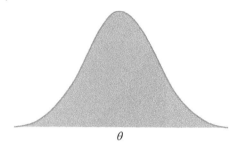

Figure 9.8 ABC step 5

It is worth noting that not all the information is captured in the summary statistics; also, the size of ϵ will have a major influence on the parameters. It has been shown as $\epsilon \to 0$, $P(\theta \mid q^*)$ converges to the true posterior distribution (Beaumont et al., 2002).

$S(\cdot)$ could be the sample mean. In cases where the actual likelihood function is unknown, it is appropriate to use different sample statistics.

9.3.2 ABC Steps

Bertorelle et al. (2010) (Figures 9.1–9.3) present an extensive overview and different steps of ABC and provide user guidelines.

- **Step 1:** Compute a summary statistic for your data.
- **Step 2:** Draw model parameters from your prior.
- **Step 3:** Simulate data from the θ draws.
- **Step 4:** Reject draws whose summary statistics are further than the tolerance (ϵ) after applying a known distribution function.
- **Step 5:** The remaining θ draws are an approximation of the posterior distribution of the model parameter.

9.4 Markov Chain Monte Carlo Application

The following analysis is part of the working paper of Galvan-Nunez and Attoh-Okine (2016).

$$\sigma_s = \theta_1 + \theta_2 t + \varepsilon, \tag{9.4}$$

where

- σ_s : Standard deviation of surface (in.)
- θ_1 : Intercept (in.)
- θ_2 : Degradation rate (in./month)
- t: Time (months)
- ε : White noise $\sim N(0, s)$

Output analysis in MCMC is important for two reasons. First, because the initial estimations about the parameters are usually characterized as bad quality, the determination of the burn-in period is critical for convergence. The burn-in is the period that corresponds to the interval of iterations starting from the first iteration in which the solutions are discarded to speed up the stationary distribution of the Markov chain. However, there is not an exact rule to determine the burn-in length, so analysis regarding this point is important. In this paper, the burn-in period was set as 5000 iterations after a trial-and-error procedure. Figures 9.9 and 9.10 show the trace and density, respectively, for parameters θ_1, θ_2, and s. With regard to the prior distribution, non-informative priors were considered. The motivation for using non-informative priors was minimizing subjective assessments that can impact the posterior distribution. As presented in the previous section, non-informative priors may lead to improper distributions. To determine the prior distribution, a lognormal and uniform distribution with parameters $\theta_1 \sim LN(0.1, 0.001)$, $\theta_2 \sim LN(0.1, 0.001)$, and $s_1 \sim U(0.0001, 1)$ were defined.

The MCMC convergence is another point to be analyzed in the MCMC output. Time series play an important role in this case, because they represent the algorithm convergence through iterations. One of the drawbacks is that the time series are autocorrelated, so the informative properties of the output are not the best. One way to maximize the information of the time series is to eliminate the autocorrelation, that is, by making each iteration independent of the posterior distribution (Figure 9.11). This strategy consists of lengthening the

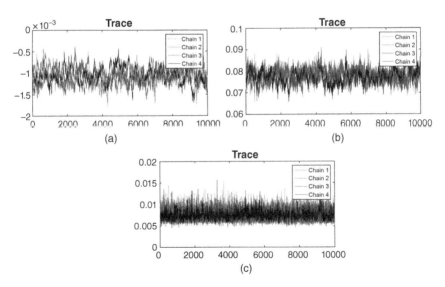

Figure 9.9 Trace. (a) Intercept, (b) degradation rate, (c) white noise

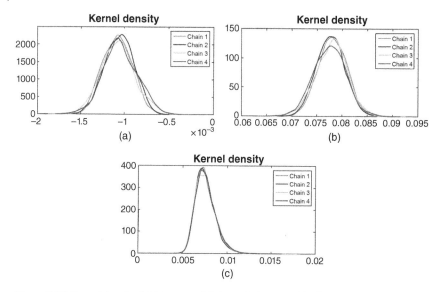

Figure 9.10 Kernel density. (a) Intercept, (b) degradation rate, (c) white noise

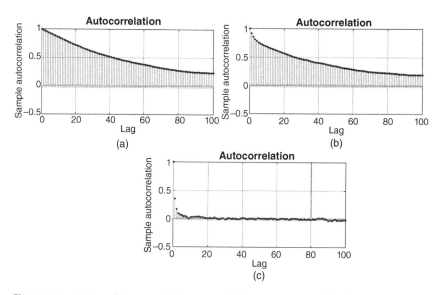

Figure 9.11 Autocorrelation plot. (a) Intercept, (b) degradation rate, (c) white noise

MCMC run by a factor m and taking every mth iteration. This process is known as thinning.

The MCMC parameters are shown as follows:

- Burn-in: 5000 iterations
- Number of samples: 10000
- Lag: 10

9.5 ABC Application

ABC, as noted in the previous sections, can be applicable to cases when the MCMC method appears to be complicated, although it is less efficient than the MCMC method. The approach is applicable to both geometry defect and rail defect data. The following section illustrates an application for determining the posterior distribution of a geometry variable, for example, cross-level.

Let assume for a prior $\theta_1 = \mu \sim U[-10, 10]$ and $\theta_2 = \sigma \sim Exp(1)$. That is, $\theta = (\theta_1, \theta_2) = (\mu, \sigma) \in \mathbb{R}^2$.

Assume that the data set consists of i.i.d values: $CL_1, CL_2, \ldots, CL_n \sim N(\mu, \sigma^2)$. If we have measured cross-level values $cl = (cl_1, cl_2, \ldots, cl_n)$ for the particular time period. The main objective is to generate samples from the posterior distribution $\theta = (\theta_1, \theta_2) = (\mu, \sigma^2)$ for a given observation of cross-level (cl).

Let us choose the summary statistic,

$$S(x) = \left(\frac{1}{n} \sum_{i=1}^{n} cl_i, \frac{1}{n} \sum_{i=1}^{n} cl_i^2 \right). \tag{9.5}$$

The process is as follows (Table 9.2)(Voss, 2013):

a) Sample $\mu_j \sim U[k_1, k_2]$, $\sigma_j \sim Exp(k_3)$
b) Sample $CL_{j,1}, \ldots, CL_{j,n} \sim N(\mu_j, \sigma_j^2)$
c) Let $S_{j,1} = \frac{1}{n} \sum_{i=1}^{n} CL_{j,i}$, and $S_{j,2} = \frac{1}{n} \sum_{i=1}^{n} CL_{j,i}^2$
d) Accept $\theta_j = (\theta_1, \theta_2) = (\mu_j, \sigma_j^2)$ if $(S_{j,1} - S_1^*)^2 + (S_{j,2} - S_2^*)^2 \leq \delta^2$, where S_j^* is based on the observed data

Figure 9.12 shows ABC simulation process.

Table 9.2 Cross-level data[a].

$CL_{1,1}$	$CL_{1,2}$	$\cdots$	$CL_{1,1000}$	$S_{1,1}, S_{1,2}$
$CL_{2,1}$	$CL_{2,2}$	$\cdots$	$CL_{2,1000}$	$S_{2,1}, S_{2,2}$
$CL_{3,1}$	$CL_{3,2}$	$\cdots$	$CL_{3,1000}$	$S_{3,1}, S_{3,2}$
$\vdots$	$\vdots$	$\cdots$	$\vdots$	$\vdots$
$CL_{30,1}$	$CL_{30,2}$	$\cdots$	$CL_{30,1000}$	$S_{30,1}, S_{30,2}$

a) Let assume $n = 300$.

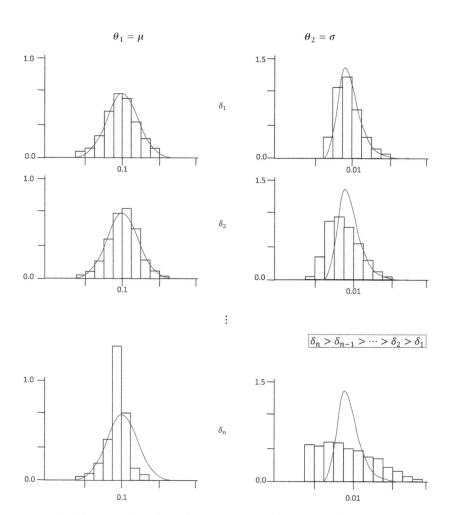

Figure 9.12 Example of ABC simulations (histograms) Adapted from: Voss (2013)

9.6 Remarks

Although few fields use Bayesian analysis, MCMC has been successfully used in track engineering applications. The use of MCMC has achieved its potential.

Table 9.3 shows selected examples of Bayesian analysis in railway track engineering applications.

ABC has major potential in track engineering applications, especially in cases where it is very difficult to construct the likelihood functions.

Table 9.3 Selected case studies of Bayesian analysis in railway track engineering.

Authors (year)	Application	Research aim	Bayesian approach
Bouillaut et al. (2008)	Rail maintenance strategy model aimed at the prevention of broken rails	Identify precisely the impact of rail flaws on safety and availability of the railway system	Dynamic Bayesian networks
Andrade and Teixeira (2011)	Track geometry degradation	Predict rail track geometry degradation	Monte Carlo simulation
Andrade and Teixeira (2012)	Track geometry degradation	Predict rail track geometry degradation	Markov chain Monte Carlo – Gibbs sampling
Andrade and Teixeira (2013)	Track geometry degradation	Predict rail track geometry degradation and thus guide planning maintenance and renewal actions	Hierarchical Bayesian models, Markov chain Monte Carlo
Lam et al. (2014)	Identification of railway ballast damage under a concrete sleeper	Address the problem of detecting railway ballast damage under a concrete sleeper	Bayesian analysis (Bayesian probabilistic approach)
Wellalage et al. (2013)	Predicting future conditions of railway bridge elements	Overcome invalid future condition prediction in existing methods	Markov chain Monte Carlo
Sinha and Feroz (2016)	Obstacle detection on railway tracks	Deployment of microelectromechanical systems (MEMS) sensors in railway track monitoring in order to detect obstacles like rock and timber dropping along the track	Monte Carlo-Bayesian analysis

References

A. R. Andrade and P. F. Teixeira. Uncertainty in rail-track geometry degradation: Lisbon-Oporto line case study. *Journal of Transportation Engineering ASCE*, **687**:193–200, 2011. doi: 10.1061/(ASCE)TE.1943-5436.0000206.

A. R. Andrade and P. F. Teixeira. A Bayesian model to assess rail track geometry degradation through its life-cycle. *Research in Transportation Economics*, **36**(1):1–8, 2012. doi: 10.1016/j.retrec.2012.03.011.

A. R. Andrade and P. F. Teixeira. Hierarchical Bayesian modelling of rail track geometry degradation. *Proceedings of the Institution of Mechanical Engineers, Part F: Journal of Rail and Rapid Transit*, **227**(4):364–375, 2013. doi: 10.1177/0954409713486619.

M. A. Beaumont, W. Zhang, and D. J. Balding. Approximate Bayesian computation in population genetics. *Genetics Society of America*, **162**(4):2025–2035, 2002. doi: 10.2307/1390807.

G. Bertorelle, A. Benazzo, and S. Mona. ABC as a flexible framework to estimate demography over space and time: some cons, many pros. *Molecular Ecology*, **19**(13):2609–2625, 2010. doi: 10.1111/j.1365-294X.2010.04690.x.

L. Bouillaut, O. Francois, P. Leray, P. Aknin, and S. Dubois. Dynamic Bayesian networks modelling maintenance strategies: prevention of broken rails. In *Proceedings of 8th World Congress on Railway Research {WCCR'08}*, 2008.

E. O. Buzbas and N. A. Rosenberg. AABC: approximate approximate Bayesian computation for inference in population-genetic models. *Theoretical Population Biology*, **99**:31–42, 2015. doi: 10.1016/j.tpb.2014.09.002.

S. Filippi. Approximate Bayesian computation, 2016. http://www.stats.ox.ac.uk/filippi/Teaching/cdt/abcLecture.pdf.

S. Galvan-Nunez and N. Attoh-Okine. Assessing uncertainty of track geometry degradation based on evolutionary Markov chain Monte Carlo. In *12th Annual Interuniversity Symposium on Infrastructure Management (AISIM12)*, 2016.

A. Gelman, A. Jakulin, M. Pittau, and Y. Su. A weakly informative default prior distribution for logistic and other regression models. *The Annals of Applied Statistics*, **2**(4):1360–1383, 2008. http://www.jstor.org/stable/30245139?seq=1#page_scan_tab_contents.

H. F. F. Lam, Q. Hu, and M. T. T. Wong. The Bayesian methodology for the detection of railway ballast damage under a concrete sleeper. *Engineering Structures*, **81**:289–301, 2014. doi: 10.1016/j.engstruct.2014.08.035.

C. P. Robert. Bayesian computational tools. *Annual Review of Statistics and Its Application*, **1**(1):153–177, 2014. doi: 10.1146/annurev-statistics-022513-115543.

D. Sinha and F. Feroz. Obstacle detection on railway tracks using vibration sensors and signal filtering. *IEEE Sensors Journal*, **16**(3):642–649, 2016. doi: 10.1109/JSEN.2015.2490247.

D. Sorensen and D. Gianola. *Likelihood, Bayesian, and MCMC Methods in Quantitative Genetics.* Springer Science & Business Media, 2002. ISBN: 0387954406. https://books.google.com/books?hl=en&lr=&id=c3ezZ3rK1HsC&pgis=1.

E. A. Suess and B. E. Trumbo. Introduction to Bayesian estimation. In *Introduction to Probability Simulation and Gibbs Sampling with R*, pages 195–218. Springer, New York, 2010. http://link.springer.com/10.1007/978-0-387-68765-0_8.

J. Voss. *An Introduction to Statistical Computing.* John Wiley & Sons, Ltd., Chichester, UK, 2013. ISBN: 9781118728048.

N. K. W. Wellalage, T. Zhang, and R. Dwight. Calibrating Markov chain - based deterioration models for predicting future conditions of railway bridge elements. *Journal of Bridge Engineering*, **20**(2):04014060, 2013. doi: 10.1061/(ASCE)BE.1943-5592.0000640.

10

Basic Bayesian Nonparametrics

10.1 General

Railway track data collection has evolved in the past few years; improved technology and the amount of data collected have made the use of classical statistical methods inappropriate in solving some data analytical issues. Therefore, the exploration of nontraditional statistical approaches to track data is important. Bayesian nonparametrics is one of the methods that appears to be promising for railway track applications.

Bayesian nonparametrics are Bayesian models in which the underlying finite dimensional random variable is replaced by a stochastic process. The major difference between the parametric models and nonparametric models is as follows. In parametric models, there are a fixed number of parameters to determine for an independent variable, for example,

$$y = a_0 + a_1 X$$
$$y = a_0 + a_1 X + a_2 X^2. \tag{10.1}$$

In the probability structure, this can be explained as follows:

$$p(x \mid \theta, D) = P(x \mid \theta). \tag{10.2}$$

θ is the finite set of parameters; given these parameters, the future prediction x is independent of the observed data. The parameters are therefore capable of capturing all the information present to know about the data for predicting the future data (Ghahramani, 2013).

The nonparametric model assumes that the data distribution cannot be defined in terms of such a finite set of parameters. Therefore, the number of free parameters grows with the amount of data; there is potentially infinite-dimensional parameter space. Only a finite subset of parameters are used in the nonparametric model to explain a finite amount of data. The model

Big Data and Differential Privacy: Analysis Strategies for Railway Track Engineering, First Edition. Nii O. Attoh-Okine.
© 2017 John Wiley & Sons, Inc. Published 2017 by John Wiley & Sons, Inc.

complexity grows with the amount of data. This is therefore a memory-based approach.

There are two broad approaches:

- Dirichlet process (DP)
- Gaussian process (GP)

10.2 Dirichlet Family

The Dirichlet distribution (DD) is the multivariate generalization of the beta distribution. It can be conceptualized as a probability distribution of probability mass functions (PMFs). This means that a single outcome is a vector of numbers. The element of this vector is between 0 and 1. A particular observation from a DD, for a railway geometry variable, is (0.2, 03, 0.5) for a case presenting a three-dimensional outcome. A four-dimensional outcome is (0.1, 0.2, 0.3, 0.4). For example, for a 3-tuple randomly drawn specific geometry defect, the probability can be shown by three different inspectors:

Inspector A: (0.15, 0.2, 0.65)
Inspector B: (0.1, 0.15, 0.75)
Inspector C: (0.33, 0.33, 0.33)

Depending on the number of inspectors, there will be lots of vectors (the only common property is that they sum up to 1). The main objective is to have the probability model that will indicate how likely each vector of belief is to appear in a sample. Furthermore, all the vectors are equally likely to occur.

Mathematically, the DD is parameterized by a vector α of positive real numbers. The probability density function (PDF) is

$$f(x_1, x_2, \ldots, x_k; \alpha_1, \alpha_2, \ldots, \alpha_k) = \frac{\Gamma\left(\sum_{i=1}^{k} \alpha_i\right)}{\prod_{i=1}^{k} \Gamma(\alpha_i)} \prod_{i=1}^{k} x_i^{\alpha_i - 1}, \tag{10.3}$$

where $\sum_{i=1}^{k} x_i = 1$ and $\{x_i\}$ is nonnegative.

Therefore, every realization of the DP is itself a PMF.

10.2.1 Moments

Let $\alpha_0 = \sum_{i=1}^{k} \alpha_i$. α_0 be called the scale:

$$E[x_i] = \frac{\alpha_i}{\alpha_0} \quad \text{mcan.} \tag{10.4}$$

$$\text{Var}[x_i] = \frac{\alpha_i(\alpha_0 - \alpha_i)}{\alpha_0^2(\alpha_0 - \alpha_i)} \quad \text{variance.} \tag{10.5}$$

10.2.1.1 Marginal Distribution

$$x_i \sim Beta(\alpha_i, \alpha_0 - \alpha_i). \tag{10.6}$$

When $k = 2$,

$$x_2 = 1 - x_1$$

$$f(x_1, x_2; \alpha_1, \alpha_2) = \frac{\Gamma(\alpha_1 - \alpha_2)}{\Gamma(\alpha_1)\Gamma(\alpha_2)} x_1^{\alpha_1 - 1}(1 - x_1)^{\alpha_2 - 1}. \tag{10.7}$$

This is the beta distribution.

10.3 Dirichlet Process

The DP is the infinite-dimensional generalization of the DDs. DP is also the continuous case of the DD. The DP includes two base parameters: a positive scalar (concentration) parameter v, which expresses the belief toward G_0, and a probability base distribution G_0. The larger the value of v, the more data is concentrated on G_0.

A generic DP can be defined as $G_{DP} \sim \text{Dirichlet}(v, G_0)$. The DP is therefore distribution over distributions. For a random measure G to be distributed according to $DP(v, G_0)$, its marginals do not follow DP. Let $A_1, \dots, A_r$ be any finite measurable partition of θ; then

$$\left(G(A_1), \dots, G(A_r)\right) \sim \text{Dirichlet}\left(vG_0(A_1), \dots, vG_0(A_r)\right) \tag{10.8}$$

using the explanation presented by Heydari et al. (2016). If it is assumed that a real line represents the entire sample space of a given parameter, the line can be partitioned into $(-\infty, A_1)(A_1, A_2) \dots (A_{n-2}, A_{n-1})(A_{n-1}, \alpha)$. Then the probability PR of falling into each interval will be as follows:

$$\begin{aligned} PR_1 &= G(A_1) \\ PR_2 &= G(A_2) - G(A_1) \\ PR_{n-1} &= G(A_{n-1}) - G(A_{n-2}) \\ PR_n &= 1 - G(A_{n-1}). \end{aligned} \tag{10.9}$$

PR_0 is the probability for the baseline distribution G_0. $PR_{0,n-1} = G_0(A_{n-1}) - G_0(A_{n-2})$.

The probabilities PR_1 to PR_n follow a DD $(PR_1, PR_2, \dots, PR_n) \sim \text{Dirichlet}$ $(vPr_{0,1}, vPr_{0,2}, \dots, vPr_{0,n})$. The distributions drawn from a DP prior have probability one, and they are discarded.

There are a number of ways to describe DPs, including stick breaking, Chinese restaurant process (CRP), polya urn, and others. This section will focus on the stick-breaking construction.

10.3.1 Stick-Breaking Construction

The stick-breaking process makes explicit the following properties of G:

- It is discrete, even if the base measure is not.
- An infinite number of different discrete distributions can be drawn (assuming the base measure is not finite).

From the baseline distribution G_0, generate a vector with random variables $\theta_0, \theta_2 - \theta_0$. $[P \sim DP(v, G_0)]$ is equivalent to

$$P = \sum_{k=1}^{\infty} \pi_k \delta_{\theta_k}, \tag{10.10}$$

where

- π_k – probability weight
- θ_k – point mass (Dirac mass) at θ

Draw the probability weight from the density function $Beta(1, v)$, and generate a vector of random variables $v_1, v_2, \ldots$, then

$$\pi_k = V_k \prod_{\ell < k} (1 - V_\ell), \tag{10.11}$$

$$\mathrm{PR}(V_n) = v V_n^{v-1}, \tag{10.12}$$

$$E(V_n) = (1 + v)^{-1}. \tag{10.13}$$

For small values v, the random breaking properties V_n will tend to be large. Assign probabilities $\mathrm{PR}_1, \mathrm{PR}_2, \ldots, \mathrm{PR}_n$

$$\begin{aligned}
\mathrm{PR}_1 &= V_1 \\
\mathrm{PR}_2 &= (1 - V_1)V_2 \\
\mathrm{PR}_3 &= (1 - V_1)(1 - V_2)V_3. \\
&\vdots \\
\mathrm{PR}_n &= (1 - V_1)(1 - V_2) \ldots (1 - V_{n-2})(1 - V_{n-1})V_n
\end{aligned} \tag{10.14}$$

The process above is analogous to breaking a stick of length 1 infinitely. Figure 10.1 illustrates the stick-breaking process.

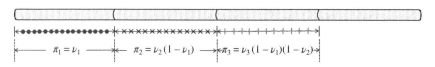

Figure 10.1 Stick-breaking process

The DP generates a discrete distribution that can be represented as a countable mixture of point masses with locations drawn independently from G_0.

The most common application of DP is in the "infinite mixture model" where the number of clusters is unknown a priori. In this situation, a DP-distributed discrete random measure is used as a prior over the parameters of the mixture component in the mixture model. Mathematically,

$$G \sim DP(\nu, G_0), \tag{10.15}$$

$$\theta_i \mid G \sim G, \tag{10.16}$$

$$x_i \mid \theta_i \sim F(\theta_i), \tag{10.17}$$

where

- x_i – observable variability
- θ_i – parameters of the mixture component

x_i belongs to F and represents the distribution of the mixture component θ_i that can be a single parameter. The generated distribution F is configured by cluster parameters θ_i, and it can be used to generate x_i observation. The density distribution is $F_X(x) = \sum_{k=1}^{\infty} \pi_k F(\cdot \mid \delta_{\theta_k^*})$, which is the mixture distribution. π_k is the mixing proportion, and the mixing components are $F(\cdot \mid \delta_{\theta_k^*})$.

10.3.2 Chinese Restaurant Process

The CRP is based on the property that $n - 1$ independent variables distributed by a probability measure generate a DP $\theta_1, \ldots, \theta_{n-1} \sim G$ and $G \sim DP(\nu, G_0)$.

The next draw from G, θ_n, has a probability greater than zero of repeating. The value of using previous draws (Granell et al., 2014). In addition, those that appear more times are more likely to appear again than those that appear fewer. This property can be explored for clustering analysis. The Chinese restaurant analysis can be used as an analogy. The tables are the clusters and the dishes on the table are the parameters of that cluster (Figures 10.2 and 10.3).

10.3.3 Chinese Restaurant Process (CRP) for Infinite Mixture

CRP is used to model the latent variable z_i of the cluster assignment. Instead of using θ_i to denote the cluster parameter and assignment, one uses z_i. In CRP, a new θ is sampled when one needs to create a new cluster.

Equation 10.18 is a generative model that describes how the data x_i and the clusters are generated.

$$\theta_{1,2,\ldots,\infty} \sim G_0$$

$$z_{1,2,\ldots,n} \sim CRP(\alpha). \tag{10.18}$$

$$x_{i=1,2,\ldots,n} \sim F(\theta_{z_i})$$

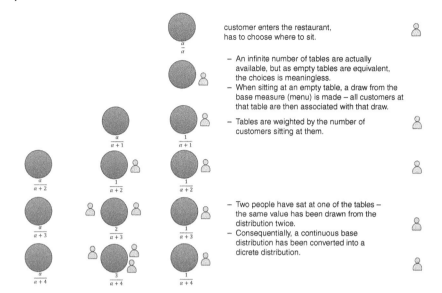

Figure 10.2 Chinese restaurant process

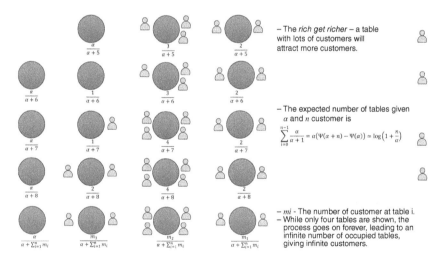

Figure 10.3 Chinese restaurant process continued

The collapsed Gibbs sampling are as follows:

- **Step 1** Initialize the z_i clusters assignment randomly.
- **Step 2** Repeat until convergence:
  - Select randomly a x_i.
  - Keep the other z_i fixed for every $j \neq i : z_{-i}$.

– Assign a new value on z_i by calculating the CRP probability that depends on z_j and x_j:

$$z_i \sim P(z_i \mid z_{-i}, X_i) \tag{10.19}$$

10.3.4 Nonparametric Clustering and Dirichlet Process

Standard clustering algorithms like k-means assume a fixed number of components, which have to be chosen a priori.

10.4 Finite Mixture Modeling

Finite mixture models with independent and identically distributed observations have PDF:

$$p(x : \theta, \pi) = \sum_{k=1}^{K} \pi_k(x; \theta_k), \tag{10.20}$$

where

- θ_k: component specific
- π_k: mixture coefficient
- $\sum_{k=1}^{K} \pi_k = 1$
- $0 \leq \pi_k \leq 1$

Finite mixture models have extensively discussed the method and approach. The section only attempts to introduce the concept and its application to railway track engineering (McLachlan and Peel, 2000).

Let us consider a two-component system: track ties. The mixture model takes the form

$$F(\vec{y}) = \lambda \sum_{j=1}^{k} F_{j1}(y_j) + (1 - \lambda) \sum_{j=1}^{k} F_{j2}(y_j), \tag{10.21}$$

where the sum of the weights is $\lambda_i = 1$, $\vec{y} = (y_1, \dots, y_k)'$ and the cdf of each component is factorized by an independence assumption. The vector y can be assumed as a vector outcomes repeat defects, whose distribution is allowed to vary depending on whether the location has a geometry defect or not. The track inspection clearly observes the defects but does not observe the geometry defects. The main objective of this problem is to nonparametrically identify and estimate the cdfs, F_{ji}, for $j = 1, \dots, k$, $i = 1, 2$.

The nonparametric Bayesian method can be used to clearly identify the model. The prior distribution is specified on the space of all possible distribution functions F. The DP is continually used.

Two parameters are used to specify the model:

a) A distribution function, F_0
b) α: scalar precision parameter – addressing the variability

For example, if the component is a Weibull function given by

$$f(x; \theta) = \left(\frac{\gamma}{\beta}\right)\left(\frac{x - \alpha}{\beta}\right)^{\gamma - 1} e^{-\left(\frac{x - \alpha}{\beta}\right)^{\gamma}},$$ (10.22)

where $\infty > x \geq \alpha, \beta > 0, \gamma > 0, \theta = (\alpha, \beta, \gamma)', F(x, \theta) = 1 - e^{-\left(\frac{x - \alpha}{\beta}\right)^{\gamma}}$.
In terms of the Dirichlet mixture model,

$$f(x, \Psi) = \lambda f_1\left(x, \theta_1\right) + (1 - \lambda)f_1\left(x - \theta_1\right),$$ (10.23)

where $\Psi = (\lambda_1 \theta_1 \theta_2)$, $\theta_i = (\alpha_i, \beta_i, \gamma_i)'$, $i = 1, 2$, $F(x, \Psi) = \lambda f_1(x, \theta_1) + (1 - \lambda)$
$f_1(x, \theta_1)$.
The normal mixture density is used when applying the EM algorithm.
For example, a mixture of Weibull distributions can be used to model tangent and curve defects:

$$f(x) = \lambda(\text{Weibull}_1) + (1 - \lambda)(\text{Weibull}_2),$$ (10.24)

where

- Weibull$_1$: Tangent defects
- Weibull$_2$: Curve defects

$$f(x) = \lambda(\text{Weibull}_1) + (1 - \lambda)(\text{Weibull}_2),$$ (10.25)

where

- Weibull$_1$: Track defects (tangent + curve)
- Weibull$_2$: Tie defects

10.5 Bayesian Nonparametric Railway Track

Mokhtarian et al. (2013) discussed how Bayesian nonparametrics can be used in the reliability analysis for railway systems at the component level. The authors highlighted the importance of Bayesian nonparametrics since most components' failure data in some railway subsystems are unknown. The authors used a mixture model of the lognormal distribution and the inverse Gaussian distribution. The authors used a simulated model to validate their results and to compute their output with a traditional parametric model:

$$f(x) = \lambda \cdot \text{LN}(a, b) + (1 - \lambda) \cdot \text{IGN}(c, d)$$ (10.26)

10.6 Remarks

The Bayesian nonparametric approach is naturally set to analyze a combination of different track defects, like ties, tie plates, and rail anchors, as a mixture of components or a missing or broken fastener algorithm based on machine vision.

Trinh et al. (2012) presented a model that focuses on solving the following problems:

a) Detect ties, tie plates, and various types of rail anchors.
b) The assessment consists of two anchors.

The authors used the Hough transform to form the basis of the analysis. The systems the authors developed can perform in real time at a vehicle speed of 10 mph at a frame rate of 20 fps. The main challenge in this approach is how to handle local shadows.

Chellappa et al. (2015) presented an extensive algorithm based on computer vision algorithms, such as

a) A crack detection based on decomposing images into edge and texture components
b) A missing/broken fastener algorithm based on computer vision
c) A crumbling/chipped tie detector based on a material classifier

The authors used a convolutional neural network to analyze objects, as well as traditional neural networks, discriminant analysis, wavelet, and support vector machine (SVM) methods.

Feng et al. (2014) used the latent Dirichlet allocation (LDA) method – a probabilistic clustering method used to cluster words into semantic topics. LDA is also a data-driven method. Noticing the disadvantages of the LDA approach, the authors used a new probabilistic structure topic model (STM) to model fasteners. This approach simultaneously learns the probabilistic representations of different objects using unlabeled samples. The STM approach makes use of a graph model representation and Gibbs sampling. It is worth noting that earlier automatic inspection detection based on classifiers is not too reliable. It also appears that previous methods have not done a good job in extracting heavy shadows before initiating the analysis. Previous studies have shown that the empirical mode decomposition in two dimensions has the capability of extracting heavy shadows.

References

R. Chellappa, X. Gibert, and V. Patel. Robust anomaly detection for vision-based inspection of railway components. Technical report DOT/FRA/ORD-15/23,

Federal Railroad Administration, 2015. https://www.fra.dot.gov/eLib/details/ L16634.

H. Feng, Z. Jiang, F. Xie, P. Yang, J. Shi, and L. Chen. Automatic fastener classification and defect detection in vision-based railway inspection systems. *IEEE Transactions on Instrumentation and Measurement*, **63**(4):877–888, 2014. doi: 10.1109/TIM.2013.2283741.

Z. Ghahramani. Bayesian non-parametrics and the probabilistic approach to modelling. *Philosophical Transactions of the Royal Society of London, Series A: Mathematical, Physical, and Engineering Sciences*, **371**:1–20, 2013. doi: 10.1098/rsta.2011.0553.

R. Granell, C. J. Axon, and D. C. H. Wallom. Impacts of raw data temporal resolution using selected clustering methods on residential electricity load profiles. *IEEE Transactions on Power Systems*, **30**(6):3217–3224, 2014. doi: 10.1109/TPWRS.2014.2377213.

S. Heydari, L. Fu, D. Lord, and B. K. Mallick. A flexible modeling approach using Dirichlet process mixtures: application to municipality-level railway grade crossing crash data. In *Transportation Research Board 95th Annual Meeting*, pages 1–18. Transportation Research Board, 2016.

G. McLachlan and D. Peel. *Finite Mixture Models, Wiley Series in Probability and Statistics*. John Wiley & Sons, Inc., Hoboken, NJ, 2000. ISBN: 9780471721185.

P. Mokhtarian, T. K. Ho, and T. Suesse. Bayesian nonparametric reliability analysis for a railway system at component level. In 2013 IEEE International Conference on Intelligent Rail Transportation Proceedings, pages 197–202, 2013. ISBN: 9781467352772.

H. Trinh, N. Haas, Y. Li, C. Otto, and S. Pankanti. Enhanced rail component detection and consolidation for rail track inspection. In *2012 IEEE Workshop on the Applications of Computer Vision (WACV)*, pages 289–295. IEEE, 2012. ISBN: 978-1-4673-0234-0. http://ieeexplore.ieee.org/lpdocs/epic03/wrapper .htm?arnumber=6163021.

11

Basic Metaheuristics

11.1   Introduction

Big data may offer opportunities and challenges in global optimization analysis. The dimensionality of the data may have a major influence on the performance of various optimization algorithms. Furthermore, most of the traditional optimization techniques may have difficulty in handling dynamic data. The data may come from different sources; hence the use of the simple objective function may not be appropriate. The presence of unstructured and semi-structured data may also lead to data that may have many features. Metaheuristics appears to be suited for the above type of problems. Figure 11.1 displays a comprehensive presentation and relationship between data science with evolutionary algorithms and swarm intelligence.

Metaheuristics can be seen as sophisticated and intuitive methods that mimic natural phenomena and explore the solution within a feasible region in order to achieve specific goals. Although those methods do not guarantee an optimal solution, they are very suitable for solving NP-hard problems, because it is possible to find good solutions in a competitive computational time compared with the exact methods of optimization. A few metaheuristic methods are genetic algorithms (GA), particle swarm optimization (PSO), ant colony optimization (ACO), tabu search (TS), artificial immune systems (AIS), and simulated annealing (SA), among others. Only the PSO method in the context of big data analytics will be discussed in this chapter.

11.1.1   Particle Swarm Optimization

PSO has been used in the analysis of large data sets. PSO is a stochastic optimization method proposed by 1995 that computationally mimics the social behavior of individuals within a group from the interaction between its members and between them and the environment in which they operate. It is a population-based stochastic algorithm modeled on the social behaviors

Big Data and Differential Privacy: Analysis Strategies for Railway Track Engineering, First Edition. Nii O. Attoh-Okine.
© 2017 John Wiley & Sons, Inc. Published 2017 by John Wiley & Sons, Inc.

Data science

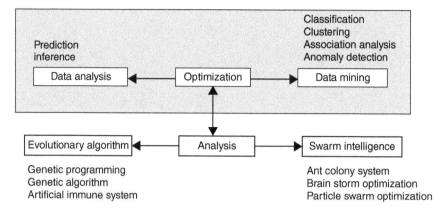

Figure 11.1 Relationship between data science with evolutionary algorithms and swarm intelligence (Cheng et al., 2016) Reproduced with the permission of BioMed Central Ltd

---

Algorithm 5 Procedure of particle swarm optimization algorithm

1: Initialize velocity and position randomly for each particle;
2: **while** the stopping criteria is not satisfied **do**
3:    Calculate each particle's fitness value
4:    Determine each particle's best position, and the best position of entire swarm
5:    **for** each particle **do**
6:      Update particle's velocity;
7:      Update particle's position;
8:    **end for**
9: **end while**

Cheng et al., 2016. Reproduced with the permission of BioMed Central Ltd

observed in flocking birds. In PSO, each particle is a potential solution in the search space, where each is associated with a fitness value, and a speed, which addresses the particles to the solution. Algorithm 5 shows the general procedure of PSO. Mathematically, a particle swarm can be represented as follows:

- N: Particle dimension
- m: Number of particles
- $X_i = (x_{i1}, x_{i2}, \ldots, x_{iN})$: Position of the ith particle
- $V_i = (v_{i1}, v_{i2}, \ldots, v_{iN})$: Velocity of the ith particle
- $P_i = (p_{i1}, p_{i2}, \ldots, p_{iN})$: Best position of the ith particle (Pbest)
- $P_{gi} = (p_{gi1}, p_{gi2}, \ldots, p_{giN})$: Best global solution of the swarm (Gbest)

During the iterative process, the particles update their positions and velocities as follows: Let $x_{il}(t)$ be the position of the particle i in the ith dimension l at time t. The particle position is updated by

$$x_{il}(t + 1) = x_{il}(t) + v_{il}(t + 1), \tag{11.1}$$

where $x_{il}(t)$ is the position of the particle i in the lth dimension at time t and $v_{il}(t)$ is the velocity of the particle i in the lth dimension at time t. In addition, the velocity is updated as

$$v_{il}(t + 1) = w(t)v_{il}(t) + (c_p r)[p_{il}(t) - x_{il}(t)] + (c_g r)[p_{gl}(t) - x_{il}(t)], \tag{11.2}$$

here, c_p and c_g are the cognitive learning and social factors, r is a random number distributed uniformly in the interval $[0, 1]$, $w(t)$ is the inertia factor at moment t, $p_{il}(t)$ is the Pbest, and $p_{gl}(t)$ is the Gbest.

11.1.2 PSO Algorithm Parameters

The parameters of PSO can have a major influence on the performance of the PSO. The parameters include the following:

a) Number of particles
b) Number of iterations
c) Velocity components
d) Acceleration coefficients

A huge amount of particles may increase the computational complexity. Similarly, a low number of iterations may prematurely stop the iteration. The velocity components tend to provide a memory of the previous flight directions. The last term in Equation 11.2, called cognitive components, measures the performance of the particles relative to past performances.

The approach of PSO in big data can be viewed from two perspectives (Cheng et al., 2013). The first approach is to lead the swarm to a certain place within the feasible region in order to develop some routine data mining, such as parameter tuning. The second approach involves moving particles that are in a low-dimensional feature space with the purpose of forming clusters; in this case, activities related to dimension reduction of the data are performed.

11.2 Remarks

There are opportunities to explore this area and to exploit optimization features for big data applications. PSO can be used in the following:

- Clustering
- Association rules
- Classification

Table 11.1 Selected examples.

Author (year)	Application	Optimization approach	Comments
Budai et al. (2006)	Track preventive maintenance scheduling	Heuristics: – Single component strategy – Most frequent work first heuristic – Opportunity-based heuristic – Most costly work first heuristic	– Objective function: minimize track possession costs and maintenance costs, given a set of routine activities and projects – The optimization model seeks to schedule the jobs together as much as possible and not necessarily as soon as possible
Zhang et al. (2013)	Track maintenance scheduling	Genetic algorithms	– Objective function: minimize the total costs, including the unsafe cost, the lost lifetime cost, the maintenance cost, and the travel cost – Factors considered: safety of transportation service, cost of useful life loss of maintained segments, maintenance cost and travel cost
Lidén (2015)	Survey of planning problems and conducted research	–	– Literature review covering more than 60 research papers regarding the use of optimization for solving track infrastructure maintenance planning problems – Little work has been conducted in the operations research area regarding infrastructure maintenance as compared to train traffic operations – Operations research has focused more on train operation problems than on infrastructure maintenance activities
Wen et al. (2016)	Scheduling tamping	Mixed integer linear programming (MILP)	– Objective function: minimize the net present cost – Technical and economic factors: track quality degradation over time, track quality thresholds based on train speed limits, impact of previous tamping operations on the track quality recovery, track geometrical alignment, tamping machine operation factors, and discount rate
Vale et al. (2010)	Scheduling tamping	Nonlinear mixed problem	– Objective function: minimize the total number of tamping actions on a track for a predefined time horizon – The paper considers the evolution over time of the track degradation, track layout, dependency of the track quality at the moment of maintenance operations, and track quality limits that depend on train speed

Metaheuristics are methods that have shown a better performance in terms of computational time rather than exact methods. Furthermore, we are interested in population-based metaheuristics because they provide more diversity in terms of potential solutions to be considered in the searching space. On the other hand, due to the fact that those methods can converge prematurely, it is important to include a sensitive analysis of the parameters of the optimization method or consider the idea of using hybrid metaheuristics methods that can work in a complementary way.

The pertinence of using metaheuristics in big data is that those methods have demonstrated a good performance in solving NP-hard problems (Toth and Vigo, 1995), such as the vehicle routing problem, scheduling problems, and others.

On the other hand, it is also relevant to point out that there are very few big data applications in railway engineering, a field in which large amounts of data are collected from sensors with the purpose of making data analysis that can prevent accidents (e.g., preventive maintenance), and traditional data mining techniques can be at a disadvantage for handling such amounts of data.

Taking into consideration what has been discussed above, we are interested in performing big data analysis supported by metaheuristics that reinforce the prediction and identification of hidden patterns in railway engineering. The next steps include the identification of the metaheuristic and its use in the big data analysis framework. In addition, the type of analysis to be performed must be developed with a clear definition of the optimization to be considered, that is, the number of objectives. Table 11.1 presents selected examples of optimization in railway track engineering.

References

G. Budai, D. Huisman, and R. Dekker. Scheduling preventive railway maintenance activities. *The Journal of the Operational Research Society*, **57**(9):1035–1044, 2006.

S. Cheng, Y. Shi, Q. Qin, and R. Bai. Swarm intelligence in big data analytics. In *Lecture Notes in Computer Science*, Volume 8206, pages 417–426, number 60975080, Springer-Verlag, Berlin Heidelberg, 2013.

S. Cheng, B. Liu, T. O. Ting, Q. Qin, Y. Shi, and K. Huang. Survey on data science with population-based algorithms. *Big Data Analytics*, **1**(1):3, 2016. doi: 10.1186/s41044-016-0003-3.

J. Kennedy and R. Eberhart. Particle swarm optimization. In Proceedings of ICNN'95 – International Conference on Neural Networks, pages 1942–1948, 1995.

T. Lidén. Railway infrastructure maintenance – a survey of planning problems and conducted research. *Transportation Research Procedia*, **10**:574–583, 2015. doi: 10.1016/j.trpro.2015.09.011.

P. Toth and D. Vigo. An exact algorithm for the capacitated shortest spanning arborescence. *Annals of Operations Research*, **61**(1):121–141, 1995. doi: 10.1007/BF02098285.

C. Vale, I. Ribeiro, and R. Calcada. Application of a maintenance model for optimizing tamping on ballasted tracks: the influence of the model constraints. In 2nd International Conference on Engineering Optimization, pages 1–8, 2010.

M. Wen, R. Li, and K. B. Salling. Optimization of preventive condition-based tamping for railway tracks. *European Journal of Operational Research*, **252**(2):455–465, 2016. doi: 10.1016/j.ejor.2016.01.024.

T. Zhang, J. Andrews, and R. Wang. Optimal scheduling of track maintenance on a railway network. *Quality and Reliability Engineering International*, **29**(2):285–297, 2013. doi: 10.1002/qre.1381.

12

Differential Privacy

12.1 General

Differential privacy (DP) is a widely accepted statistical framework for protecting data privacy. DP attempts to learn as much as possible about a group while investigating or studying as little as possible about individual components of that population. Imagine a railway agency has two otherwise identical databases: one with critical track information and one with critical track information related to safety or other issues. If there is a related database where the critical or other safety issues are available, differential privacy will ensure the probability that a statistical query will produce a given result that can be the same whether it is conducted with the first or second database. The following challenges may affect the correct implementation of differential privacy:

a) The more information you "ask" of the initial database, the more noise has to be injected in order to minimize the loss of privacy. This shows the trade-off between accuracy and privacy.
b) Once the data has been leaked, it is gone (Green, 2016).

The idea of DP was initially introduced to human private data, mainly medical and health records, but it is now clear that the same ideas can have a major contribution in infrastructure systems, including railway track engineering. The major objective of this chapter is to introduce the concept of differential privacy to railway track data collection and evaluation. Big data that are not properly processed can cause various damages to the privacy issues of the data, especially in cases where multiple parties are involved, for example, track defects and railway derailments and accidents. A key point is that an adversary should not be capable of getting additional information or distinguishing between two databases based on the output.

For a classic example, let us assume a railway track data table, which has 1000 records, and each record has attributes, railway (geometry defect – cross level, gage), railway defect (TDD, cracks), and tonnage in terms of MGT. Let us assume the cracks attribute has domain size of 3; for the attribute of cross level, let us assume yes or no, and MGT should be in the range of 1200 to 1750 MGT. The following questions can be asked: How many locations have a higher cross level compared to the threshold value? In how many locations are there cross levels above the threshold in different MGT ranges? Different examples involve numerical readings from different sensors used in railway geometry measurements that are monitoring various defects. Again, questions can be as follows: given the large number of sensor readings and values across a particular subdivision, what is the average reading of a given geometry feature? How many geometrical defects are correlated with each other? How are two geometry features correlated to each other? Is it possible to predict one of the geometry features from another? These are some of the questions differential privacy will attempt to address in track engineering applications.

Currently, railway agencies are processing and querying large track data through the cloud computing platform. This may offer vast advantages, but it also means some risks in managing the data and using it for further analysis is imminent. To avoid some of these issues of cloud computing, agencies may employ encryption and other data protection methods. Differential privacy can be used to protect the privacy of these data.

By deploying various sensors and using geometry cars, and other data collection vehicles, a huge amount of real time data can be collected and reported to various control centers, for timely monitoring, control and additional analysis. Using the basic theory and ideas from DP, an attempt is made to introduce the concept to railway track data, an example in the area of civil infrastructure systems. This approach presents a new narrative in the area of infrastructure data collection, and hence railway track data collection. The application of DP will have a major influence in track condition data and rail tank safety. For example, in rail tank safety analysis, it is always difficult for different players, railway agencies, insurance companies, carriers, and legal practitioners to share information. This makes it very important that track data collection be presented so as to protect third parties in deducing some vital information, using auxiliary documents like contract information and thereby protecting the condition and vulnerability of the asset. The use of DP fits into a new paradigm of urban science where three classes of data are collected and analyzed; these include (a) the infrastructure, (b) the environment, and (c) the people.

12.2 Differential Privacy

Differential privacy DP was first introduced by Dwork (2006) as a method that guarantees that the privacy of an individual will be protected even if an adversary possesses some auxiliary background information about that individual. DP is the current application that will be discussed in the realm of sensitivity of railway track data and in addressing how the sensitivity attributes in track data can cause unwarranted danger to the safety and smooth operation of the train, for example, how an intruder to railway track monitoring data could be capable of identifying some of the critical and vulnerable parts of the track. Furthermore, how the use of auxiliary information like contract documents can also provide additional knowledge about the track conditions and vulnerability will also be discussed.

There are three kinds of privacy protections:

- Data collection protection
- Data releasing protection
- Data analysis protection

Various methods have been used to address privacy; these include (a) addition of noise and execution of anonymous schemes and (b) k-anonymity and l-diversity approaches used for data protection.

12.2.1 Differential Privacy: Hypothetical Track Application

Most track geometry parameters are collected in the form of time series. Furthermore, the data values change over time and location; there are also trends and, in some cases, periodicity. Railway agencies may need to keep their confidentiality in the data mining process since these characteristics may be sensitive information. Hong et al. (2013) highlighted the following:

a) The amplitude of time series data indicates the strength of the signal.
b) Peaks and troughs may disclose extreme changes.
c) By observing trends in time series data, an adversary may predict the future changes in the time series.
d) Periodicity will give major information about the periodic changes in the time series data.

Addition of random noise may hide the sensitivity of the data. But, again, random noise can be removed by traditional filtering methods.

Suppose we have a track geometry data set, consisting of data points

$$D = \{X_1, \dots, X_n\}, \tag{12.1}$$

where X_i can be a geometry variable of a selected location. This location may have some important information about the condition and subsurface condition of the site. Now consider an algorithm K that takes a database D as input and outputs h. This output will in most cases provide information about the site (location). Under the differential privacy paradigm to the data, the data will be treated as follows. Let D and D' differ by one element. The algorithm k satisfies

ϵ different privacy for sets S and adjacent databases $D \sim D^\sim$. $\Pr(\cdot)$ is defined as the privacy exposure risk.

$$\Pr(K(D)\epsilon S) \leq e^\epsilon (PK(D^\sim)) \in S. \tag{12.2}$$

Intuitively, let x be a location of the geometry variable in the database D, and D' contains the same information as D except x's data are replaced by a default data. The differential privacy guarantees that the probability of any output of the mechanism K is within e^ϵ multiplicative factor whether or not x's sensitive information was included in the input. The parameter ϵ controls how much the distribution of outputs can depend on the data from the selected location x.

Differential privacy:

$$e^{-\epsilon} \leq \frac{\Pr(K(D)) \in S}{\Pr(K(D^\sim)) \in S} \leq e^\epsilon, \tag{12.3}$$

where S is in Range(K) and ϵ is privacy metric.

The epsilon-differential privacy can be explained as follows: let us assume a learner implements summary statistics called $K()$, an adversary proposes, two data sets D and D', that may differ by only one row and a test Q.

$K()$ is called epsilon-differentially private if

$$\left| \log \frac{\mathrm{Prob}[K(D) \; in \; Q]}{\mathrm{Prob}[K(D') \; in \; Q]} \right| \leq \epsilon \tag{12.4}$$

for all adversary choices of D, D', and Q. Q can be an interval. For example, let $D - \{0, 0, 0, \dots, \theta\} - \{1000\text{zeros}\}$ be binary values of defects per meter at the particular track segment (0 – no defect). $D' - \{1, 0, 0, \dots, \theta\} - \{999\text{zeros}\}$ 1 – is the presence of one defect above the threshold. The set Q will be an interval $[T, 1]$ where the adversary picks the threshold T. The adversary's goal is to pick T such that $K(d) \leq T$. The learner has two goals:

a) To pick an algorithm $K()$ such that $K(D)$ and $K(D')$ are close enough that the adversary is unable to reliably get T
b) To have $K()$ be a good estimate of the expectation

Let the mean be $m(.) = K(D) = 0$ when evaluating D, $K(D') = 0.001$ when evaluating D'. Therefore, if the adversary picks $T = \frac{1}{2000} = 0.0005$, the adversary can reliably obtain $K(D)$ having evaluated D'.

$$\Pr(K(D) \in S) \leq e^\epsilon \Pr(K(D^\sim) \in S) + \delta, \tag{12.5}$$

where ϵ and δ are parameters that control the trade-off between false alarm (type I) and missed detection (type II) errors.

For the formally sensitivity function ΔK (Figure 12.1), the worst case difference that the DP algorithm for a practical algorithm K will have to hide in order to achieve the presence or absence of a selected input can be defined as

$$\Delta K = \max_{D, D^\sim} \| K(D) - K(D^\sim) \| . \tag{12.6}$$

Figure 12.1 Sensitivity function

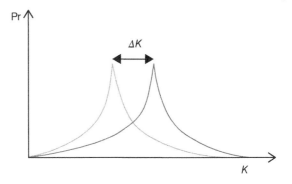

D and $D^\sim$ differ by one element at most, and addition of noise in the form of a Laplace distribution results in

$$\Pr[K(D) + r = k] \le e^\epsilon \Pr[K(D^\sim) + r = k], \tag{12.7}$$

where r is randomly chosen noise.

To use differential privacy, the following steps are as employed (Bambauer et al., 2013)

The noise is sampled from a Laplace distribution into a probability density function

$$\text{Laplace}(x, \lambda) = \frac{1}{2\lambda} e^{-|x|/\lambda}, \tag{12.8}$$

where λ is determined by both ΔK and the desired privacy coefficient parameter ϵ.

1) Select ϵ. The smaller the value, the greater the privacy.
2) Compute the response to the query using the original data. Let a represent the true answer to the query.
3) Compute the global sensitivity (ΔK) for the query.
4) Generate a random value (noise) from a Laplace distribution with mean $= 0$ and scale parameter $b = \Delta K/\epsilon$.
5) Provide the user with response $= a + y$. The noise added (y) is unrelated to the characteristics of the actual query (number of observations in the database or query and the value of the true response and is determined exclusively by ΔK) and ϵ.

Figure 12.2 presents the general structure of differential privacy.

Figure 12.3 shows how DP can be useful in rail track accidents involving hazardous materials.

The differential privacy practically can be applied by estimating the mean of the samples. Let us assume that there are N subdivisions, each with m different attributes of unknown effect:

$$x_{ij} = \mu + z_{ij},$$

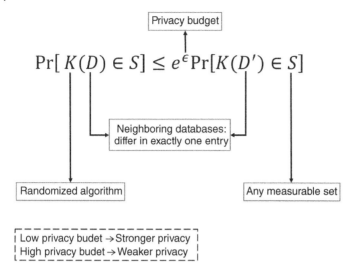

Figure 12.2 General structure of differential privacy

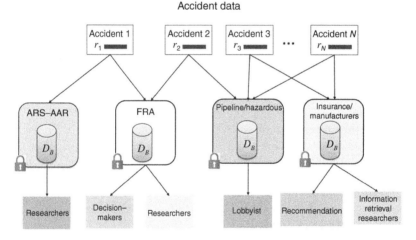

Figure 12.3 An example of DP in rail tank safety

$i = 1, 2, \ldots, N$ and $j = 1, 2, \ldots, m$ where μ is an unknown mean and z_{ij} is normally distributed. The sample mean for each subdivision can be calculated as follows:

$$\widetilde{X} = \frac{1}{m} \sum_{j=1}^{m} x_{ij} = \mu + \frac{1}{m} \sum_{j=1}^{m} z_{ij}.$$

The sample $\widetilde{X}$ has a mean of zero and a variance of $\frac{1}{m}$.

The ϵ-differential privacy estimate of μ is

$$\widetilde{X} = \frac{1}{m} \sum_{j=1}^{m} x_{ij} + \frac{1}{\epsilon m} w_i,$$

where w_i is a Laplace random variable with unit variable (Sarwate et al., 2014). Again, differential privacy has been used in different regression analysis settings (Fredrikson et al., 2014). The idea can be extended to classification analysis.

12.3   Remarks

Railway agencies collect vast amounts of data about individual tracks and locations that may contain sensitive information about the system. The information, once in the wrong hands, can constitute both safety and legal issues for the track agencies. Therefore, a key challenge is how the railway agencies will design and develop analysis techniques from the large-scale data while at the same time protecting key information in this process. Although the initial differential privacy was applied to medical privacy, the incident of cybersecurity breaches makes the application recommended to be applied in railway track and other railway applications.

References

J. R. Bambauer, K. Muralidhar, and R. Sarathy. Fool's Gold: An Illustrated Critique of Differential Privacy. *Vanderbilt J. Entertain. Technol. Law*, vol. **16**, no. 4, p. Paper No. 13–47, 2013.

C. Dwork. Differential privacy. In *Automata, Languages and Programming*, pages 1–12. Springer, Berlin Heidelberg, 2006. http://link.springer.com/10.1007/11787006_1.

M. Fredrikson, E. Lantz, S. Jha, S. Lin, and D. Page. Privacy in pharmacogenetics: An end-to-end case study of personalized warfarin dosing. In 23rd USENIX Security, pages 17–23, 2014. https://www.usenix.org/node/184490.

M. Green. What is Differential Privacy? -A Few Thoughts on Cryptographic Engineering, 2016. https://blog.cryptographyengineering.com/2016/06/15/what-is-differential-privacy/.

S. K. Hong, K. Gurjar, H. S. Kim, and Y. S. Moon. A survey on privacy preserving time-series data mining. In 3rd International Conference on Intelligent Computational Systems (ICICS'2013), pages 44–48, 2013. http://psrcentre.org/images/extraimages/10413598.pdf.

A. D. Sarwate, S. M. Plis, J. A. Turner, M. R. Arbabshirani, and V. D. Calhoun. Sharing privacy-sensitive access to neuroimaging and genetics data: a review and preliminary validation. *Frontiers in Neuroinformatics*, **8**(2):35, 2014. doi: 10.3389/fninf.2014.00035.

Index

Big Data and Differential Privacy: Analysis Strategies for Railway Track Engineering, First Edition. Nii O. Attoh-Okine.
© 2017 John Wiley & Sons, Inc. Published 2017 by John Wiley & Sons, Inc.

Wiley Series in
Operations Research and Management Science

Operations Research and Management Science (ORMS) is a broad, interdisciplinary branch of applied mathematics concerned with improving the quality of decisions and processes and is a major component of the global modern movement towards the use of advanced analytics in industry and scientific research. The *Wiley Series in Operations Research and Management Science* features a broad collection of books that meet the varied needs of researchers, practitioners, policy makers, and students who use or need to improve their use of analytics. Reflecting the wide range of current research within the ORMS community, the Series encompasses application, methodology, and theory and provides coverage of both classical and cutting edge ORMS concepts and developments. Written by recognized international experts in the field, this collection is appropriate for students as well as professionals from private and public sectors including industry, government, and nonprofit organization who are interested in ORMS at a technical level. The Series is comprised of four sections: Analytics; Decision and Risk Analysis; Optimization Models; and Stochastic Models.

Advisory Editors • Analytics
Jennifer Bachner, Johns Hopkins University
Khim Yong Goh, National University of Singapore

Founding Series Editor
James J. Cochran, University of Alabama

Analytics
Yang and Lee • *Healthcare Analytics: From Data to Knowledge to Healthcare Improvement*
Attoh-Okine • *Big Data and Differential Privacy: Analysis Strategies for Railway Track Engineering*

Forthcoming Titles
Kong and Zhang • *Decision Analytics and Optimization in Disease Prevention and Treatment*

Decision and Risk Analysis
Barron • *Game Theory: An Introduction,* Second Edition
Brailsford, Churilov, and Dangerfield • *Discrete-Event Simulation and System Dynamics for Management Decision Making*
Johnson, Keisler, Solak, Turcotte, Bayram, and Drew • *Decision Science for Housing and Community Development: Localized and Evidence-Based Responses to Distressed Housing and Blighted Communities*
Mislick and Nussbaum • *Cost Estimation: Methods and Tools*